SUSTAINABLE TRANSPORT, MOBILITY MANAGEMENT AND TRAVEL PLANS

Transport and Mobility Series

Series Editors: Professor Brian Graham, Professor of Human Geography, University of Ulster, UK and Richard Knowles, Professor of Transport Geography, University of Salford, UK, on behalf of the Royal Geographical Society (with the Institute of British Geographers) Transport Geography Research Group (TGRG).

The inception of this series marks a major resurgence of geographical research into transport and mobility. Reflecting the dynamic relationships between socio-spatial behaviour and change, it acts as a forum for cutting-edge research into transport and mobility, and for innovative and decisive debates on the formulation and repercussions of transport policy making.

Also in the series

Integrating Seaports and Trade Corridors
Edited by Peter Hall, Robert J. McCalla, Claude Comtois and Brian Slack
ISBN 978 1 4094 0400 2

International Business Travel in the Global Economy
Edited by Jonathan V. Beaverstock, Ben Derudder, James Faulconbridge and Frank Witlox
ISBN 978 0 7546 7942 4

Ports in Proximity
Competition and Coordination among Adjacent Seaports
Edited by Theo Notteboom, César Ducruet and Peter de Langen
ISBN 978 0 7546 7688 1

Railways, Urban Development and Town Planning in Britain: 1948–2008
Russell Haywood
ISBN 978 0 7546 7392 7

Transit Oriented Development
Making it Happen
Edited by Carey Curtis, John L. Renne and Luca Bertolini
ISBN 978 0 7546 7315 6

For further information about this series, please visit www.ashgate.com

Sustainable Transport, Mobility Management and Travel Plans

MARCUS ENOCH
Loughborough University, UK

ASHGATE

Published by
Ashgate Publishing Limited
Wey Court East
Union Road
Farnham
Surrey, GU9 7PT
England

Ashgate Publishing Company
Suite 420
101 Cherry Street
Burlington
VT 05401-4405
USA

www.ashgate.com

British Library Cataloguing in Publication Data
Enoch, Marcus.
Sustainable transport, mobility management and travel plans. -- (Transport and mobility series)
1. Business travel--Planning. 2. Business travel--Environmental aspects. 3. Social responsibility of business.
I. Title II. Series
658.4'083-dc23

Library of Congress Cataloging-in-Publication Data
Enoch, Marcus.
Sustainable transport, mobility management and travel plans / by Marcus Enoch.
p. cm. -- (Transport and mobility)
Includes bibliographical references and index.
ISBN 978-0-7546-7939-4 (hardback) -- ISBN 978-0-7546-9860-9 (ebook)
1. Urban transportation--Planning. 2. Transportation demand management. 3. Choice of transportation. 4. Urban transportation policy.
I. Title.
HE305.E56 2011
388.4--dc23

2011041809

ISBN 9780754679394 (hbk)
ISBN 9780754698609 (ebk)

MIX
Paper from responsible sources
FSC® C018575
www.fsc.org

Printed and bound in Great Britain by the MPG Books Group, UK.

List of Figures

Contents

List of Tables

Preface

The Transport Challenge

Local transport systems worldwide are facing significant challenges. In particular, whilst increased car ownership and use has generated huge benefits to many individuals in society it has also led to an array of difficulties. These include increased congestion and a deteriorating environment in urban areas which in turn lead to more dispersed development patterns in suburban and rural areas that become ever more difficult to serve by public transport, thus pushing more people to own and use a car further exacerbating the problems. This situation is important to address because the transport system is integral in allowing people, organisations and nation states to function.

The Travel Plan Solution

As defined later in this book, Travel Plans comprise a means of delivering measures to help facilitate sustainable transport through an additional 'mediating institution'.

Otherwise known as (among other things) site-based mobility management plans, employer transport plans, commuter plans and green transport plans, from the perspective of some local authorities they can be an attractive mechanism for delivering transport improvements. This is because they can be quick, cheap and effective to introduce, whilst also being politically acceptable. By contrast, traditional approaches to addressing transport issues tend to have significant downsides. Thus, building new roads or railways can be effective but is expensive, time-consuming and disruptive; improving public transport services is acceptable but expensive and has only a marginal impact on car use; and restricting or charging for car use can be cheap, quick and effective but is politically unpopular.

These strengths of the Travel Plan concept arise because of two key differences from other transport policy instruments. First, rather than being a single policy instrument, the Travel Plan is actually a delivery mechanism of a package of several other policy instruments, which means that a blend of measures can be tailored to the specific needs of a site so as to offer a balanced strategy that can ensure that car use is sufficiently controlled, but that incentives are accurately targeted enough to maintain the acceptability of the plan to the users. Second, Travel Plans are implemented not by the local authority, but typically by organisations such as employers. This has two key effects:

- Travel Plans are developed effectively at the neighbourhood level and consequently focus directly on the transport needs of the users in that local area;
- Travel Plans replace the (often negative) relationship between local authority and citizen with a more positive relationship (for example between employer and employee or between school and parent/pupil).

Crucially then, Travel Plans work by developing highly user-focused balanced packages of incentives and disincentives in a partnership that involves local government, transport providers, organisations and perhaps most importantly, individual users. In theory, if done properly, Travel Plans offer transport and wider benefits to individuals (through more travel choices); to implementing organisations (via cost savings, happier and healthier staff and better company image); to providers (through additional business opportunities); and to government (congestion reduction, improved air quality).

Yet despite these strengths and all round benefits the take up of Travel Plans, even in countries where they have been known for a decade or more, is relatively low. Such a state of affairs is clearly sub-optimal and essentially derives from the (lack of) acceptance by transport generating organisations of their role in addressing transport problems.

Aim

Accordingly, this book introduces the Travel Plan concept and then provides a detailed examination of how it has evolved over time. Practical and policy recommendations are offered as to how Travel Plans should be developed in the future. In particular, the book focuses on the relationships between government, transport providers, the transport generating organisations and the individual users, and seeks to draw lessons as to how these interactions might be improved.

Method

The book is based on a review of the existing literature, supplemented by a number of face-to-face interviews and a series of email and telephone communications to provide orientation (Bogner et al 2009).

The interviews were conducted by the author over the course of a decade of involvement in research into sustainable transport policy in general, and Travel Plan policy in particular for a range of clients including the European Commission, UK National Government Departments and UK regional and local authorities (see the end of this Chapter for a list of these projects). Specifically, the majority of interviews conducted were in-depth, semi-structured and exploratory, which allowed a degree of flexibility enable the interviewees to elaborate where necessary and even to develop new themes whilst maintaining control when necessary (Drever 1995). Meanwhile, the question topic areas for the interviews were developed based on an initial review of the literature and of course were heavily dependent on the nature of the research project being conducted at that time.

Meanwhile, the email and telephone 'personal communications' referred to were far shorter and more focused, and sought to either:

- answer specific questions about what had happened or was happening to Travel Plans at a particular time and place; or
- speculate on what was likely to happen to Travel Planning in the future.

In contrast to the interviews, most of these were conducted during the course of the last three years specifically for the purpose of this book.

In terms of those interviewed and contacted, these included Travel Plan practitioners (typically Travel Planners, consultants or service providers) and policy officers at the local, regional, national and European level who were generally involved in specific Travel Plan initiatives which were relevant to one or other of the research projects. Also interviewed were several 'experts', that is people who process special knowledge of a social phenomenon which the interviewer is interested in (Gläser and Laudel 2004), in this case generally commentators on the Travel Plan sector including academics, journalists and consultants. In this research the experts were selected on the basis of their experience with in-depth knowledge of the sector, and the experiences and insights of these experts were felt to be useful in teasing out the key issues surrounding Travel Planning. Accordingly, those participants could be said to have been selected based on a purposive sampling strategy, whereby interviewees/respondents are selected to meet the specific needs of the researcher (Trochim 2008; Miles and Huberman 1994).

In order to gain an understanding of the Travel Plan sector, data were then analysed using thematic analysis, which Boyatzis (1998: 4) suggests involves developing an explicit code, which "may be a list of themes, a complex model with themes, indicators and qualification that are casually related or something in between these two forms". Or, in other words, thematic analysis is a process that determines commonalities, relationships and differences across a set of different responses (Gibson and Brown 2009).

Structure

The book is structured as follows. Chapter 1 outlines the issues and challenges facing transport planners and policies, while Chapter 2 establishes the Travel Plan concept and charts its development. Chapters 3, 4, 5 and 6 then investigate how Travel Plans have evolved in terms of four 'axes of development', namely segment, scale, scope, structure and support. Chapter 7 then presents a series of recommendations for practitioners and policy makers to support the development of the Travel Plan, and then concludes by examining the current status of Travel Plans around the world and considering the future prospects of the Travel Plan.

References

Bogner, A., Littig, B. and Menz, W. 2009. *Interviewing Experts*, (London: Palgrave Macmillan).

Boyatzis, R.E. 1998. *Transforming Qualitative Information: Thematic Analysis and Code Development*, (London: Sage Publications).

Drever, E. 1995. *Using Semi-structured Interviews in Small-scale Research: A Teachers Guide*, (Edinburgh: Scottish Council for Research in Education).

Gibson, W.J. and Brown, A. 2009. *Working with Qualitative Data*, (London: Sage Publications).

Gläser, J. and Laudel, G. 2004. *Experteninterviews und Qualitative Inhaltsanalyse*, (Wiesbaden: VS Verlag für Sozialwissenschafen).

Miles, M.B. and Huberman, A.M. 1994. *Qualitative Data Analysis*, 2nd Edition, (Thousand Oaks: Sage Publications).

Trochim, W.M.K. 2006. *Research Methods Knowledge Base – Non-probability Sampling*, Web Center for Social Research. [Online] Available at http://www.socialresearchmethods.net/kb/sampnon.htm [accessed 8 July 2010].

Related Travel Plan projects

Travel Reduction Attainment Via Energy-efficient Localities PLANning (TRAVEL PLAN PLUS) project, European Commission, Executive Agency for Competitiveness and Innovation, Intelligent Energy Europe, STEER Programme, Brussels, 2008–2011.

A possible blueprint for mainstreaming Travel Plans in the UK? Project for the UK Department for Transport and the National Business Travel Network, London, 2007–2008.

Best practice guide for implementing local Travel Plan groups for Transport for London, London, 2006.

OPTIMUM2 project for the Interreg IIIB programme of NWE, European Commission, Lille, 2004–2007.

Taxation futures for sustainable mobility for the Economic and Social Science Research Council, London, 2003.

Opportunities for mobility management, a series of small scale studies for the London Borough of Southwark, London, 2002–2003.

Evaluating the effectiveness of the site specific advice programme for Travel Plans for the UK Department for Transport, London, 2002–2003.

Evaluation of government departments' Travel Plans for the UK Department of Transport, Local Government and the Regions, London, 2001.

The impact of the personal tax regime on modal choice for the UK Department of the Environment, Transport and the Regions, and UK Inland Revenue, London, 2000–2001.

Contributing to a sustainable integrated transport system for the Scottish Executive, Glasgow, 1999–2000.

Fair and efficient pricing in transport – the role of charges and taxes for the European Commission, Brussels, 1999–2000.

Acknowledgements

I would like to take this opportunity to thank everyone who contributed to the process of creating this book. In particular, first I'd like to thank those who gave their time to offer me their perspectives on Travel Plans. Second, I'm most grateful to a number of undergraduate, MSc and PhD students here at Loughborough University who have worked hard on Travel Plan-related projects in recent years and so helped me to better understand the topic area. Third, I thank my colleagues for their help. Especially, researchers Lian Zhang and Lisa Davison have provided much support, as has fellow geek day buddy Lucy. Next I'd like to thank Vicky Eves and Ken Chatterton for their help in realising many of the illustrations; Angela Underhill for proofing the draft manuscript in such a short time; and the publishing team at Ashgate, particularly Katy Crossan and Matt Irving.

Finally, I would like to give my biggest thanks to my parents and my brother David for their encouragement over the years; to Stephen Potter, James Warren and Stephen Ison for their helpful advice; and most of all to my wife Sudaxshina, without whose initial inspiration and subsequent good humour (most of the time) this book could not have been completed.

Please note that copyright for all of the plates (photographs) rests with the author, as it does for all of the figures apart from the first (Figure 1.1) – to which I acknowledge the *Island Studies Journal* for granting me permission to publish it here. Data for the tables is sourced where appropriate on a case by case basis. It is also politic to note that much of Chapter 4 is based on data from the OPTIMUM2 and TRAVEL PLAN PLUS European Union-funded projects, and that Chapter 5 emerged from a joint National Business Travel Network and UK Department for Transport study about the future of Travel Plans.

Marcus Enoch
March 2012

To SP, who encouraged me to write

To SI, who encouraged me to write a book

To SM, who encouraged me to write this book

Chapter 1
Transport Issues and Their Solutions

1 Defining Transport

According to Hoyle and Knowles (1998: 1), transport exists to "provide for the movement of people [and] goods and for the provision and distribution of services … Transport thereby fulfils one of the most important functions and is one of the most pervasive activities in any society or economy". Faulks (1977: 16) adds that the movement of passengers or goods should "occur from where they are to where they would prefer to be or to where their relative value is greater" – that is that the action should have a purpose. And Cole (2005: 5) notes "transport is a service rarely in demand for its own characteristics. Demand [for transport] is usually derived from some other function".[1] Two further key elements that make transport what it is, are the spatial and temporal dimensions involved.

Thus, transport can be said to involve movement of people and goods, be purposeful, and (generally) only occur as a consequence of other activities taking place in different locations at different times.

1.1 The Importance of Transport

In terms of the importance of transport, Vuchic (2007: 1) notes that "the founding, shaping and growth of human agglomerations throughout history have been products of complex interactions of many forces. One major force has always been transportation …". He continues that transport specifically has had a major role in determining the locations of cities, their size, urban form and urban structure. On similar lines, Headicar (2009: 1) begins his book with the phrase "transport is a vital part of everyday life" and Shaw et al (2008: 3) states that "the importance of travel and transport to the functioning of our economies and societies is hardly in doubt, but the very ordinariness of transport systems often means that they are taken for granted".

Transport then, is a subject worthy of further study. And as good a place to start as any, is with the transport system – that is the means of effecting transport.

---

1 It should be noted that a growing body of research contests this perspective of transport being merely a derived demand by suggesting that communication on the move by wireless email, mobile phones and social networking can make travelling a work and/or leisure time in itself. For more information on this, see Knowles et al (2008) and the September 2011 special issue of the *Journal of Transport Geography* on Geographies of the Passenger (Adey et al 2011).

1.2 The Transport System

> The physical and social worlds are full of change, paradox and contradiction. There is no simple unchanging stable state … Systems are patterned, regular and rule-bound, but through their workings they can generate unpredictable features and unintended effects (Dennis and Urry 2009: 48).

Such a comment is especially appropriate for transport, which although relatively easy to simplify – at a basic level the transport system can be said to comprise of fixed elements or 'infrastructure' and mobile elements or 'vehicles', that interact with each other in compliance of a series of rules – in actuality is extremely complex.

This complexity stems from the fact that the transport system does not exist in isolation (almost by definition as noted earlier). It is therefore appropriate to set out some of the factors that influence the transport system and in turn at how it impacts on other sectors.

2 Influences on the Transport System

Before this step is made though, it is helpful to refer to a visual representation of a highly simplified series of interactions made from a transport planning perspective (see Figure 1.1), which provides some idea at least of how these elements interact with each other.

Specifically, a range of political, economic, social, technological, legal and environmental factors combine to influence the demand for transport in an area. These 'demand factors' can be grouped into 'sub-systems'. For example, the urban sub-system comprises activity patterns, transport infrastructure and land use related factors, while the car costs sub-system includes information relating to the cost of motoring. The next step occurs where the sub-systems then influence the organisation and operation of transport elements such as car ownership levels, number of trips, length of trips, level of public transport use and the interaction of these then generates the resultant transport-related impacts (for example, personal mobility, congestion, noise, air pollution).

2.1 The 'Unique Demand Profile'

A slightly different way of trying to model the interactions within the transport and the wider system is to think of every square kilometre of land as being unique in terms of the amount and type of transport it needs.

On choosing one of these 'Unique Demand Profiles', each rather like a fingerprint, the next step is to break it into three separate components (see Figure 1.2). The first of these is the natural or physical terrain (or topography) of the location. This is the 'baseboard', or the land with no human contribution. The second component consists of the human development (houses, shops, factories, schools and so on) which form the 'demand' generating activity centres. The final component comprises roads, railways, airports and vehicles of the transport system – that is the 'supply'. Put these three components together and then connect them to other similar UDPs in the form of a jigsaw puzzle, and one has a transport system.

Developing this model further, it is possible to see how the roles of the transport operator, the transport planner and the transport policy maker might interconnect (see Figure 1.3). In brief, the transport policy maker sets out the values, objectives, constraints and rules of engagement to inform the task of the transport planner, who in turn directly influences the levels of demand (and hence the operating conditions) of the transport operator. Feedback on the performance of the system is then passed from the transport operator to the transport planner and from the transport planner to the transport policy maker to complete the loops.

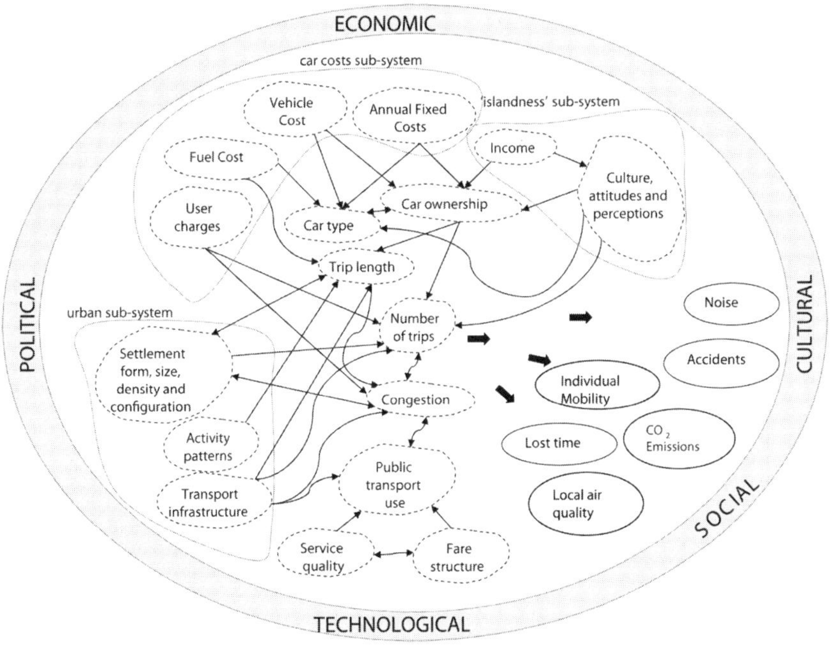

Figure 1.1 Influences on the transport system
Source: Warren and Enoch (2010). Reprinted with permission

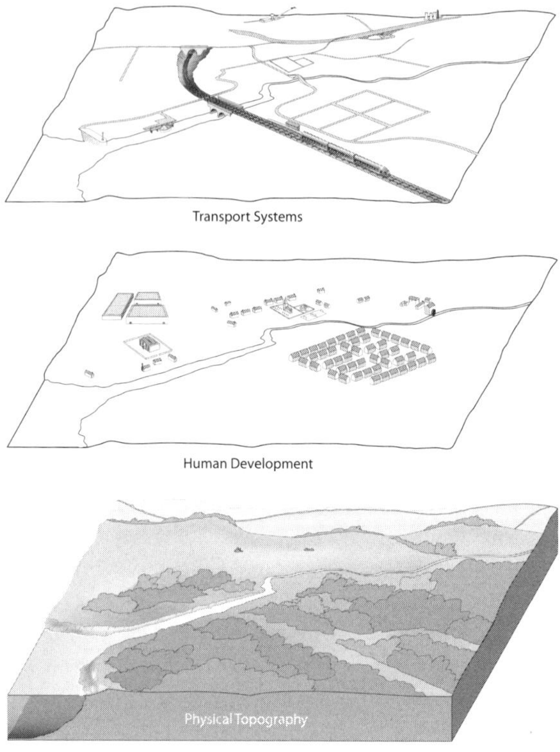

Figure 1.2 Unique demand profiles and transport

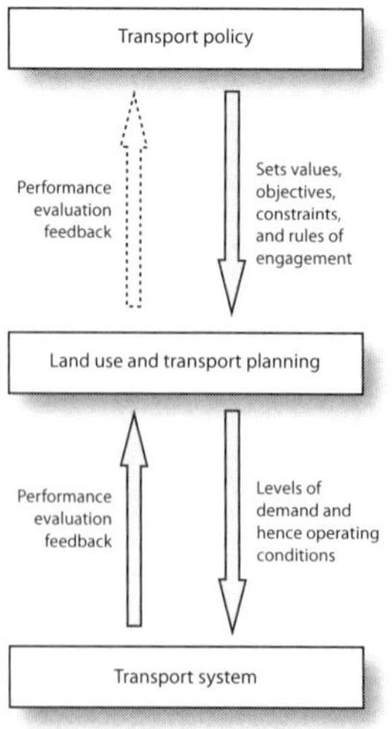

Figure 1.3 Stakeholder roles in transport

2.2 Travel Demand and Supply Characteristics

Understanding the Unique Demand Profile concept is a start, but in addition there are several important features relating to the nature of the demand and the supply of transport which help explain why the topic is so difficult to deal with. Thus, Ortuzar and Willumsen (1994) notes the special features of travel demand are:

- First, the demand for transport is *qualitative* (that is lots of attributes are not quantifiable).
- Second, it is *differentiated* (by time of day, day of week, journey purpose, importance of speed and frequency). This means that transport demand is very difficult to analyse and forecast and that any service without these attributes being matched may well be useless.
- Third, transport is a *derived* demand, not an end in itself. This means that we must understand how facilities to satisfy these human needs (work, leisure, health and so on) are distributed over time and space. A good transport system widens opportunities for people to undertake these activities, while a congested and/or poorly connected system restricts options and limits economic and social development.
- Fourth, transport demand is heavily governed by *space* – this means it is the distribution of activities over space which makes for transport demand and so in the vast majority of cases space must be explicitly considered. The most common way of approaching this is to divide study areas into zones and to code them together with transport networks. This spatiality of demand often leads to problems where, for example, there are too many people and not enough taxis in one area and too many taxis and not enough customers in another.
- Also important is the *dynamic* element of transport. So, a good deal of transport demand is concentrated on a few hours each day, in particular in urban areas where most of the congestion takes place during peak periods. This time variable nature once more adds a further layer of complexity.

Meanwhile, the same book reports that the supply characteristics are:

- Transport supply is a *service* and not a good. In other words, transport cannot be stored (so as to save it up for times of peak demand) but must be consumed when produced, otherwise benefit is lost. This is why it is so important to predict demand as accurately as possible so as to use resources as efficiently as possible.
- Many of the characteristics of a transport system derive from being a service. In broad terms, a transport system consists of a number of fixed assets (*infrastructure*) and a number of mobile units (*vehicles*). The movement of goods and people becomes possible when these are combined together with a set of rules for their operation.

- Further complexity is caused in that the supply of transport is shared among a whole range of government departments and agencies, developers, businesses, operators and travellers. This has economic implications too, in that travellers often do not pay the actual cost of using the transport system meaning it is not used as efficiently as it might be.
- Next, the provision of transport supply is lumpy – in other words one cannot provide half a runway, or a third of a bus. It is often therefore difficult to match the supply with the demand because of this. Related to this is that investments in transport infrastructure are not only lumpy, but can take many years to carry out. Furthermore, disruption during the construction phase involves additional costs to users and non-users alike.

2.3 Balancing Demand and Supply

Fundamentally, the task of the transport planner is straightforward. It is to match the demand for transport by society with the supply or so-called capacity of the transport system in a given place at a given time. Yet as has just been noted, while simple to articulate, this is complicated to do. Consequently it is maybe surprising just how often the 'ideal' situation where demand and supply are in balance (or something close to it) occurs. In many locations and for much of the time though, either the level of transport demand is too high for the amount of available capacity (supply), or else the demand is not sufficient to make efficient use of the supply.

The next section explores these two cases in a little more detail.

2.4 Where Transport Demand Outstrips Supply

Where (and when) there is too much demand for the available supply it can be seen that a series of impacts result at a series of deepening levels (see Table 1.1 for an illustrative representation of the point made in this section).

At the initial direct level, public transport vehicles become overcrowded whilst road networks become congested. Such effects lead to second-order issues whereby discomfort and (often unpredictable) delays result. In addition, noise and local pollutants (carbon monoxide, sulphur dioxides, particulates) increase, as does the amount of energy and carbon dioxide used, and road traffic 'accidents' become more likely. Subsequent third-order effects at the individual level see delays and noise, for example leading to stress with implications for mental health (and by extension family relationships); whilst local pollutants contribute to pulmonary-related health conditions such as asthma. Meanwhile at the community level, aggregated delays in the transport system (made worse by the uncertainty) alongside the worsening health effects reduces productivity and increased energy use further increases costs, thus negatively affecting the economy. And social cohesion is damaged as children no longer play on the streets and people see less of their neighbours due to rising levels of traffic.

Table 1.1 Impacts where transport demand exceeds supply

Level of impact	Type of impact	Impact diffusion
Direct impacts	Public transport overcrowding worsens	
	Traffic congestion rises	
Second-order impacts	Unpredictability and amount of delays increase	
	Discomfort increases	
	Noise increases	
	Local pollutants increase	
	Energy use increases	
	Carbon dioxide increases	
	Increased stress levels	
Third-order impacts	Increased mental health issues	
	Increased pulmonary health complaints	
	Decreased productivity	
	Increased energy costs for industry	
	Reduced social cohesion	
	Increased obesity in children	

2.5 Where Transport Supply Exceeds Demand

In the case where there is too little demand for the capacity provided, the causality chain operates in a different manner to that described in the previous section although there are also some similarities. Here, where road space is increased the behavioural impulses are for people to drive faster, to make longer trips more often and at times more convenient to them (typically in the peak), which in turn implies that development patterns become more dispersed leading to greater overall vehicle kilometres travelled and reductions in public transport and non-motorised trips. Following on from this, one might expect there to be more road traffic accidents, increased energy use and carbon dioxide emissions. One might also expect poorer health – both in terms of higher levels of obesity in adults and particularly children due to less physical activity, and in mental health from people being forced to travel further to live their daily lives. Meanwhile, the damage to social cohesion by reduced social interaction is this time caused by the health effects just referred to together with severance effects (where additional road or rail infrastructure coupled with higher traffic speeds). Table 1.2 provides a diagrammatic representation for this section.

Table 1.2 Impacts where transport supply exceeds demand

Level of impact	Type of impact	Diffusion of impact
Direct impacts	Faster driving	
	Longer and more frequent trips	
Second-order impacts	More dispersed development	
	Reduced use of public transport, walking and cycling	
	Noise increases	
	Local pollutants increase	
	Energy use increases	
	Carbon dioxide increases	
	Increased stress levels	
Third-order impacts	Increased mental health issues	
	Increased pulmonary health complaints	
	Decreased productivity	
	Increased energy costs for industry	
	Reduced social cohesion due to new infrastructure and faster and higher levels of car use severing communities	
	Increased obesity in adults and children	

However, perhaps the main issue where and when supply outstrips demand concerns the wastage of resources (such as transport capacity, capital investment, operating subsidy, time, land, energy and especially political capital). This is a crucial point in a resource constrained world, and means that the environment whereby demand outstrips supply is becoming increasingly dominant.

2.6 Quantifying these Effects

In attempting to quantify these impacts, one of the most recent and comprehensive sources of evidence is the *External costs of transport update study*, which reports cost estimates for the year 2000 using statistics for 17 European Union States (EU-17) (Schreyer et al 2004). This separates the congestion effects from the remainder and reports figures for three congestion-related approaches, thus:

- If welfare theory is used to take account of costs arising from the inefficient use of existing infrastructure, a figure of €63 billion is derived (just over 0.5 per cent of GDP);

- If the optimal congestion charging amount was levied, revenues would be €753 billion (almost 8.5 per cent of GDP); and
- If loss of time was the indicator the cost would be €268 billion (3 per cent of GDP).

Meanwhile, the total external costs caused by transport activity on the rest of society excluding congestion amounts to €650 billion, or 7 per cent of total GDP. Climate change accounted for 30 per cent of this, while air pollution, accidents, noise and nature/landscape/urban impacts took up 27 per cent, 24 per cent, 7 per cent and 5 per cent respectively. In terms of mode, road transport dominates with 84 per cent of the impacts, aviation generates 14 per cent and rail and waterways together cause the remainder. Finally, passenger transport contributes two thirds of the costs and freight transport the rest. Regarding the level of investment in transport systems, OECD and ITF (2010) notes that typically Western European nations have spent approximately 1 per cent of GDP on inland transport systems (approximately €90 billion in 2000), of which close to 80 per cent was used for roads.

Given the scale and seriousness of these impacts then, one might expect at some point in the process to force individuals (and by extension decision makers) to take mitigation actions. These fourth-order impacts might involve people altering their time of travel, their destination, choice of mode or even suppressing their trips altogether as making them becomes too difficult. However, (clearly) this does not always occur.

2.7 Transport Trends

Such evidence is sobering enough. Yet things look even worse when one considers the trends over time. In simple terms, car ownership levels and car use figures can be used as proxy indicators for the issues discussed previously, and these are positively frightening. Specifically, Sperling and Gordon (2009) point out that currently 85 per cent of the world's population does not own a car, but that the majority aspires to do so. It goes on to predict that car use will continue to grow by 3 per cent a year and that by 2020 there will be two billion vehicles globally, at least half of which will be cars.

3 Addressing Transport Issues

In such a non-ideal world, it is the job of politicians and their officials to decide where the optimal position lies between the minimum acceptable quality of service to transport users, and the maximum level of investment in the transport system acceptable to society as a whole (and voters/tax payers in particular).

Perhaps unsurprisingly, here again there are yet more barriers. Taking a UK perspective, Shaw and Docherty (2008) applies a PESTLE analysis, and finds that political factors for non-action include the lack of political will, the short-termist, incremental political culture, complex and fragmented governance structures, civil service risk aversion, the power of transport and other vested interest groups and the attitudes and electoral importance of the middle classes. In addition, it reports a range of other economic, social, technological, environmental and legal factors including an ideological fixation with competition, evolving lifestyle trends, the lack of an apparent technical fix, local air quality issues and constitutional concerns to do with devolution issues. Similarly, in the opinion of the author while from an economic, social, environmental and a common sense perspective, 'do nothing' is not an option, from a political perspective – where the decisions are taken – do nothing may be the only option. This is for a number of reasons, but in the UK context (assumed by its residents to be particularly poorly served by transport) these include:

- Transport improvements are expensive. HM Treasury and Infrastructure UK (2010) commits the UK Government to spending £GBP30 billion (€34 billion) on transport improvements over a five-year period, with the hope that up to £GBP200 billion (€225 billion) altogether can be leveraged from the public and private sectors – that is more than £GBP3,000 (€3,400) per person in the UK or double that once non-tax payers are removed from the equation. Moreover, even local measures are costly – capital costs alone for a new single carriageway bypass are in the region of £GBP3,000 (€3,400) per metre.
- Transport measures are time consuming to introduce. The planning process is extremely long and expensive, as is the subsequent implementation stage. Thus, some UK road schemes first planned in the 1930s are only now being built. This is clearly not attractive to politicians who typically prefer demonstrable improvements within a four or five year term of office.
- Worse, the implementation of transport measures can be a dirty, messy, disruptive, noisy and thoroughly unpleasant business – in particular in the short-term and for people living close by. Moreover, while transport schemes can result in 'winners', they also result in 'losers'. Typically, while there are more winners than losers, the winners tend to gain relatively little (for example, 15 minutes per day in time savings – which is not likely to influence whether or how they vote), while losers may lose their homes and/or livelihoods. As a result, potential losers are strongly motivated to fight against any proposed schemes (thus adding to the time and costs) and to vote against any local politician they feel may have caused them to lose out.
- And, tying these threads together is the fact that transport improvements are most often needed in the very areas where each of the above elements – cost, time required and number of people adversely affected – are at their highest.

- Another issue is that there is a shortage of skills in the transport planning sector. This is partly due to a culture of the UK Government not spending money for many years since the 1970s, and partly because the skills that were formerly required (for example, designing road junctions, installing traffic signals and planning public transport networks) are relatively less important in the new world of 'smarter choices'.
- Meanwhile the fact that 'transport is a derived demand' means that it is often a complicated problem to address and is correspondingly difficult to explain.
- Finally, perhaps the most cost effective way of addressing the transport crisis is to restrict people's mobility – yet politically this is extremely unpopular.

Overall, the crux of the problem appears to be that the benefits of car use are very evident to individuals, whereas the problems are more diffuse, hit others rather than the car user, and impinge on future rather than current generations. It causes changes that cannot easily be perceived and which are accommodated on a piecemeal basis. This means that any individual action produces no obvious personal benefit. For example, one mother letting her children walk to school will do little to improve their health or safety, so long as no similar action is taken by other parents. This unequal conflict between choosing immediate and tangible personal benefit over a delayed and far less visible cost to society is behind many of the difficulties faced when addressing the transport crisis.

Transport policy, therefore, is not an area favoured by politicians, especially when compared to sectors such as education and health due to the large sums of money required, the time scales needed and their controversial nature (particularly where schemes are under construction). Moreover, pulling the planning and policy strands together, Wright and Ashford (1998) summarises that formulating a transport plan is further complicated by three factors:

- First, the problem is neither isolated nor independent.
- Second, urban transportation systems themselves form part of the overall regional and national transport system – meaning that appropriate transport planning requires an examination at various levels since policy decisions at one level may have significant impacts on the planning process and the final transportation plan.
- Third, when implemented a transport plan will affect its own environment thus modifying the demand generated and possibly invalidating the criteria on which the plan was initially based.

So, how might this difficult situation be turned around?

3.1 Transport Planning and Policy Instruments

OECD (1994) notes that congestion management measures (and transport measures more broadly) can be characterised as being either 'supply-side' or 'demand-side' measures, whereby supply-side measures aim to increase existing capacity of the transport system to improve the traffic flow of all modes and where demand-side measures are designed to reduce the level of (primarily car) demand. These can be represented as in Table 1.3.

Table 1.3 Supply-side and demand-side approaches explained

	Supply	**Demand**
Supply-side	Capacity increased or decreased in line with demand	Assumes demand 'independent' and rises or falls due to external factors
Demand-side	Assumes supply is 'fixed' – that is too difficult to manipulate	Levels of demand 'managed' to better suit available capacity

4 Supply-Side Solutions

The most obvious way of increasing the capacity is to provide more of it; either through building more infrastructure or operating more vehicles.

4.1 Building More Capacity

Regarding infrastructure, there are certainly some cases where this is the most sensible option – for example to service new developments, where small towns or villages require a road bypass, or where pinch-points severely curtail rail capacity (for example at Welwyn Viaduct to the north of London in the UK on the East Coast Main Line, where four rail tracks are reduced to two tracks to cross the River Mimram). As well as roads and railways, such measures could include pedestrian facilities, cycle routes, guided bus infrastructure, light railways, terminals and interchanges (for public transport services), park-and-ride stations, and car parks (IHT 1997).

But, there are a number of practical problems with providing new infrastructural capacity. First, it is time consuming to provide due to the complexity of the transport problem; the various legal processes; the need for lots of people to be consulted, and so on. Secondly, it is expensive and finance is required over a number of years while the scheme is built. Thirdly, such major work always results in some people losing out during the construction phase and/or the operating phase, and so schemes can be unpopular with certain affected groups. Finally, the biggest need for new capacity tends to be in areas where land is in very short supply. This means that routes for new roads and railways take even longer to work out, are even more expensive to build and upset even more people than they would otherwise.

Operating more vehicles (cars, taxis, buses, trains, aircraft) too can be sensible in certain circumstances. Once again though, cost (particularly staff costs associated with drivers and other operational staff) is usually a significant issue, whilst in addition the scope for providing higher levels of service by operating additional vehicles is often limited by infrastructural constraints which then need to be addressed.

4.2 Utilising Capacity More Efficiently

Usually rather less controversial, expensive and time consuming than constructing new transport capacity, is to find ways of more efficiently utilising existing transport capacity. There are two complementary ways of achieving this: one is to allow more vehicles to use the infrastructure, and the other is to increase the occupancy of each of those vehicles.

Instruments designed to enable the more efficient use of infrastructure by vehicles are many and varied. However, they can be basically categorised as either trying to alter/influence the spatial or the temporal (time) dimension.

On the spatial dimension side, capacity can be increased by narrowing lanes (so that three lanes can be put in place instead of two previously), by implementing one way systems in certain suitable neighbourhoods, and by modifying the geometry at junctions (often where the capacity is most constrained) to allow more traffic to pass through. Meanwhile, for the future, there are a number of so-called 'intelligent vehicle' systems being developed (particularly in the USA), whereby the 'human factor' is gradually being reduced to allow vehicles travelling at high speed to travel (safely) far closer together.

On the temporal or 'dynamic' side, supply-side options tend to be focused on solving the so-called 'peak problem' (as noted earlier). This has been addressed by transport planners in a number of ways. Thus, in some towns and cities where heavy traffic flows are observed in one direction in the morning, and in the other in the afternoon, then tidal flows lanes can be operated – that is lanes are switched on a daily basis to match the road capacity to those flows. Second, dynamic Urban Traffic Control (UTC) systems have been developed whereby traffic signals at junctions are operated according to the level of traffic approaching each leg of the intersection, therefore allowing the most efficient throughput of traffic. Next, various differential pricing schemes have been used all over the world on buses, trains and aeroplanes to try and encourage people to travel during the off-peak instead of the peak (that is discounts are offered during the off-peak). On the roads too, the congestion charging concept envisages that higher tolls are paid during busy periods than when roads are not operating at or near capacity, while parking charges are often varied according to length of stay in order to encourage short-term parking in shopping centres for example.

In summary, two core principles in increasing the efficiency of transport networks emerge. First, the quicker and smoother the traffic flow, the higher the capacity, and second, the slower the flow, the more likely it is to be smooth. From these statements, arise a few indicators as to a way forward:

- Reduce the number of 'interruptions' or 'conflicts' – hence grade separated junctions, and 'no right turns', one way streets, barriers preventing pedestrians crossing streets, subways and footbridges, caravans, lorries and coaches not allowed in the outside lanes, 'no stopping allowed' and so on.
- Reduce the 'unpredictable human' element – hence the increased use of technology in everything from UTC systems, to ramp metering on motorways, to cruise control (and ultimately Intelligent Vehicle Highway Systems) and so on.
- Spread out the peak – the closer the road/rail system is to capacity, the more unstable it is.

Increasing, vehicle occupancy/utilisation is another approach for increasing the efficiency of the transport system. For instance, in the UK the average number of people travelling in a car in 2008 was 1.6, although 60 per cent of car trips contain only one passenger (a figure which rises to 84 per cent of commuting and business trips) (DfT 2010). From several perspectives (emissions, energy used, congestion) this is not desirable from a public policy perspective.

Many mechanisms have been tried to redress this situation. The most direct has been to encourage people to share cars, that is 'lift-share' or 'carpool'. Outside a few cases where this has worked successfully (for example through informal arrangements between neighbours or through company-sponsored schemes) however, such schemes have not yet tended to work well (mainly due to people having incompatible journeys (different origins/destinations and different start/finish times) and personal security issues, though there are now indications that this is starting to change. Nevertheless, ways for transport planners to encourage lift sharing include the provision of high occupancy vehicle lanes and priority spaces for car sharers at shopping or workplace destinations, and charging more for car use, for example through road tolls, or higher fuel taxes or parking charges.

Far more common an approach to increase occupancy, has been to encourage people to use public transport, which almost by definition is of higher occupancy than a car. Once again, though, similar issues that prevent people car sharing arise – personal security, unsuitable routes/times of operation – even when mitigated with bus lanes, bus priority at junctions, cheaper fares relative to the cost of driving and so on.

Finally, in addition to the vehicle *occupancy* point, there is the issue of vehicle *utilisation*. Currently, again in the UK, there are more than 30 million cars on the road, which on average are parked for more than 22 hours a day. A current 'best guess' of those involved in the parking sector in terms of the total number of parking spaces in the UK is that the real number of parking spaces equates to around four or five spaces per car (Moran, M. 2003. Personal communication,

Editor, *Parking Review*, London, 18 October), or in the region of 120–150 million spaces, which would seem to equate to more than 1 per cent of the land area of England, Scotland and Wales. Such figures suggest that rather than owning a car, it may be rather more efficient if people could use a car only when it is needed – and hence the concept of 'car sharing' or 'car clubs' – which is a similar concept to renting a car – is beginning to take off in countries around the world.

In conclusion, increasing the capacity of the transport system to cope with the increased demand has generally been seen as the first choice solution to the transport problem. But as noted in the previous section, despite all the engineering and technological advancements, transport problems are steadily getting worse rather than improving. The reason seems to be, that even if such an approach was economically, socially, environmentally, physically and politically possible (and/or desirable), there is evidence to suggest that theoretically it is ultimately self-defeating. Most notably, the UK Government's Standing Advisory Committee on Trunk Road Assessment concluded in a 1994 report that providing more capacity 'induced more traffic' – that is providing more capacity in itself stimulates more traffic (SACTRA 1994). More graphically, Richard Moe, head of the US National Trust for Historic Preservation described this phenomenon thus: "building more roads to ease traffic is kind of like trying to cure obesity by loosening the belt" (Ball 2004).

If this is the case, the next approach is surely to try and reduce the demand for transport.

5 Demand-side Solutions

Of course demand-side solutions to transport problems have nearly always been in evidence. However, it is probably only in the last 40 years that they have been accepted as not being only a 'second best' option to expanding capacity to be used when resources were not available as before. Thus, Schiller et al (2010) suggests that 'sustainable transport' (perhaps a more positive and pro-active term for demand-side transport management) emerged from:

- Concerns about the impacts on local environments of infrastructure provision from the 1970s;
- A recognition that schemes designed to reduce traffic flows in cities often had a wide range of benefits; and
- The growth of awareness around the 'sustainability' debate following the publication of the Brundtland Report in 1987.

In addition in the UK in the 1990s there was the issue of induced traffic, alluded to earlier, which provided an intellectual basis for considering alternatives to providing more road capacity, crucially coupled with an economic recession that effectively removed the means of paying for such a policy (Parkhurst and Dudley 2008).

5.1 Demand-side Measures

There are a number of ways to reduce demand for transport, and the available measures can be categorised in a number of ways. For example, May et al (2004) divides these into 'strategies' and 'instruments' (see Table 1.4). The (re-ordered) four strategies refer to the approach they use and are: to improve road network; improve alternative modes; reduce car use; and reduce the need to travel.[2] Meanwhile, the classification of policy instruments is distinguished by how they work. There are four (again re-ordered) categories, namely:

- information (advice, awareness, communication);
- technology (new infrastructure, new and improved vehicles);
- economic (charges, taxation); and
- regulation (management, control, services).

Table 1.4 Instruments and strategies for managing travel demand

	Improve road network	Improve alternative modes	Reduce car use	Reduce need to travel
Information: advice, awareness, communication	Driver route guidance, safety guidance, traffic information provision	Real time passenger information	Awareness campaigns, individualised marketing,	Teleworking, video-conferencing
Technology: infrastructure, vehicles	Improved quality roads and parking facilities	Improved quality infrastructure and vehicles	Community roads	Transit oriented development
Economic: charges, taxation, subsidy	Road pricing, parking charges	Fares policy	Road pricing Fuel taxes Vehicle taxes	Land taxes
Regulation: management, control, service	Traffic management Urban traffic control	Bus priorities Improved service standards	Access permits Parking restrictions Traffic calming	Land use regulations Suburban-isation control

Source: Adapted from May et al (2004)

2 Note: May et al (2004) also includes a fifth strategy 'improve vehicles and fuels', but this has been ignored here as it does not aim to influence vehicle use (though of course it does affect some of the resultant impacts). Also, the table has been modified to incorporate demand-*side* rather than demand *management* measures.

From this, it can be seen that the decisions of the 'traveller' become increasingly strategic from left to right – from which route to take, to when to travel and by what mode, to whether to make the journey at all – that is 'ultimate' demand management. Similarly, the mechanisms used to manage demand become (arguably) more difficult to implement but also more effective from the top row of Table 1.4 to the bottom.

5.2 Issues with Demand-side Approaches

Reducing the need to travel is still rather a nebulous concept to grasp for most people, but in the longer term is really the only viable solution. This is problematic as people and society as a whole have adapted to having the 'right' to mobility due to all the factors described earlier (more dispersed activities, higher incomes, cheaper travel and so on). In brief, at a strategic level the evidence seems to show that:

- land use needs to be more controlled in order to allow people to walk, cycle and use public transport to access activities wherever possible;
- transport prices need to fully reflect the true cost that it imposes on society (that is they need to rise significantly);
- economic and transport growth are strongly linked – one (socially undesirable) method of reducing transport demand would be to reduce the level of economic activity;
- rationing is another way whereby demand of a product can be limited. For instance, a tradable permit scheme for aircraft emissions across Europe is currently being investigated.

More tactically, there is scope for individuals to reduce their travel needs by travelling 'smarter'. For instance, instead of people travelling five days to the office, perhaps they could work four slightly longer days or else work one day a week at home. Or, people might learn to 'trip chain', whereby people might stop at the supermarket on the way home from work instead of making a separate return trip after dinner. However, here too there are (sometimes major) attitudinal, behavioural and cultural obstacles to overcome that make such simple sounding solutions rather more difficult to introduce than it might appear. Moreover, when these are coupled with broader institutional and political pressures pushing in the opposite direction, then the take up of this type of proposal becomes even less likely.

In taking account of these facts, Steiner (1992) draws on experience from the utility sector to highlight issues with the implementation of demand-side approaches in transport, and finds that:

- There are a large number of participants acting within an uncoordinated decision-making process.
- The pricing structure is fractured and prices are not directly related to demand nor quality of service.

- The focus of demand-side activities is very narrow, responding to a shortage of capacity in specific areas at specific times. Demand-side measures are not directly related to supply-side measures.
- Technological improvements almost exclusively focused on the supply-side.

From these insights Steiner (1992) concludes that:

- Congestion pricing gives people the right incentives to shift people from cars but that political and equity considerations will make implementation difficult.
- People hope for a technological fix to solve air pollution and congestion but this is only likely in the long-term.
- For demand-side efforts to be meaningful, they need to be broad-based in the objectives they hope to achieve, in the timescales involved (short-term versus long-term), and in terms of the stakeholders and geographical areas covered. In other words, demand-side measures need to be enacted uniformly across a region, balance short-term and long-term needs and address congestion outside of the peak periods. This is, she acknowledges, not likely to be an easy task.

There would also seem to be a whole new approach to dealing with transport issues and problems (Schiller et al 2010), perhaps most particularly to do with the conflicting attitudes to the positive nature or otherwise of car use (see Table 1.5).

Table 1.5 Features of the supply-side and demand-side approaches

Supply-side	Demand-side
Emphasises mobility and quantity (more, faster)	Emphasises accessibility and quality (closer, better)
Emphasises one mode (uni-modality, automobility)	Emphasises plurality (multi-modality)
Often lacks good connections between modes	Emphasises interconnections (intermodality)
Accommodates and accepts trends	Seeks to interrupt and reverse harmful trends
Plans and builds based on forecasts of likely demand (predict and provide)	Works backwards from a preferred vision to planning and provision (deliberate and decide)
Expands roads to respond to travel demand	Manages transportation or mobility demand
Ignores many social and environmental costs	Incorporates full costs within planning and provision
Transportation planning often in 'silos' disconnected from environmental, social and other planning areas	Emphasises integrated planning combining transportation with other relevant areas

Source: Schiller et al (2010)

In addition, there is one further crucial difference to those illustrated in Table 1.5. This is that the skills needed for designing and implementing demand-side, as opposed to supply-side, transport solutions are very different. Specifically, skills focused on marketing, communication and business management are required, rather than abilities focused on engineering, operations and planning – and consequently this means that potentially different people are involved in the process who may not necessarily understand the needs and perspectives of their new colleagues. This leads us nicely to the current situation facing UK transport policy, that of the two 'transport tribes'.

6 The Two 'Transport Tribes'

In considering whether to adopt a primarily supply-side or demand-side approach to addressing the transport 'crisis', in the UK at least (and perhaps in North America and Western Europe as well) there seems to have been almost a philosophical paralysis of which to choose. This is interesting because in party political terms the transport sector has tended to be a rather pragmatic area of policy where ideological issues – generally over the 'proper' level of public versus private ownership (nationalisation versus privatisation), and control (regulation versus deregulation) – surface only from time to time. Over the last 20 years though, it has become clear that this new philosophical battleground has emerged, and in the last five years the two sides have become increasingly polarised with some significant impacts on the decision-making process.

In simple terms and at their extremes, two 'camps' have emerged.

First, the supply-side camp (transport operators, construction firms, transport users and their representatives) advocates building transport capacity wherever and whenever possible to facilitate people's lifestyle choices and improve their quality of life. Transport and mobility are seen as being a 'good thing' because they stimulate social and economic interaction by increasing accessibility and opportunity. The 'predict and providers' tend to play on the economic and social benefits additional transport capacity generates at the regional and national levels.

The second camp of mobility managers on the other hand consists of those who feel the 'predict and provide' ethos is economically and socially, as well as environmentally unsustainable due to the phenomenon of 'induced traffic' – specifically that building more capacity is ultimately self-defeating because it inherently increases demand in itself. This, they argue, is a problem because transport is not always a 'good thing' because of the wasted time, uncertainty, stress and pollution. Instead, people would often probably prefer not to travel vast distances to access services if there was an attractive enough alternative located nearby. Here, the environmental and social concerns on local communities are emphasised.

This split (or at least the size of it) is new. Historically, this area was not so philosophically controversial. Instead, as noted earlier, when demand grew and where resources (land, labour, money, political capital) allowed, then additional road, rail and air capacity would be provided. When resources did not, for example during the fuel crises of the 1970s, then a demand management approach would be applied – sometimes as a deliberate policy, sometimes not.

This changed in the UK in 1988 – earlier in the USA (see Meyer 1999 for more on this) – with the publication of two reports by the UK Department of Transport. One announced a large-scale programme of road building while the other predicted that car use would increase by 83 per cent to 142 per cent by 2025 (DoT 1988a; DoT 1988b). Tellingly, subsequent analysis demonstrated that even the significantly expanded roads programme proposed, costing £GBP6 billion (€6.8 billion) (today worth £GBP10 billion (€13 billion)) would only be able to marginally increase capacity meaning that congestion could still be expected to worsen significantly (Goodwin et al 1991). Consequently, with the publication of the 1998 White Paper *A new deal for transport* a decade later (DETR 1998), there was almost a universal consensus that demand management measures offered the obvious way forward. Unfortunately, though, this opportunity for action was missed due to the low political priority afforded to transport in the late 1990s. Ever since, the transport policy decision-making process has effectively been paralysed and perhaps the worst case of a 'do-nothing' approach has emerged by default, a situation that has at least partly arisen because of the increasingly divergent views being provided to the UK Government by exponents of the two conflicting points of view.

6.1 The Need for a New Approach

Ironically, at the city, level the transport strategies of the more progressive (at least in a transport sense) public authorities around the world have been developing 'packages' of measures containing both disincentives for car users or 'sticks' (for effectiveness), and incentives to use alternatives to the car or 'carrots' (for acceptability) since the 1970s and with some success.

For example, the Government in Singapore has been able to implement a packaged approach to transport policy over several decades apparently due in effect to a social contract between the state and the people whereby the population has generally accepted relatively limited political freedom in return for a steady improvement in the country's economic prosperity and their quality of life. Specifically, strong land use planning controls coupled with stringent car restraint measures (road pricing, parking and vehicle ownership restrictions) have been implemented alongside the development of a high quality public transport network (Willoughby 2001). Meanwhile, packages combining elements of traffic restraint and transport capacity enhancements were also promoted in cities as diverse as Gothenburg (Sweden), Groningen (Netherlands), Hong Kong, Bremen (Germany) and Lyon (France) (May 1986). And May (1991) reports on the UK case, whereby such an approach finally began to gather pace in the late 1980s with high profile studies being undertaken in Birmingham, Leeds, Bristol and Edinburgh.

Effectively, the so-called 'Package Approach' (also known as the 'Balanced Transport Strategy' or the 'Carrot and Stick Approach') has come about largely because the 'traditional' options available have significant flaws for politicians. Specifically, they can be perceived as being either disruptive, long-term and expensive (building new roads and railways); politically unpopular (road pricing and parking restrictions) or else expensive and relatively ineffective at reducing car use (enhancing public transport services). In other words, in the grand scheme of things, neither of the 'two tribes' has an exclusive grasp on the 'correct' way forward.

Moreover, there is also some evidence that Dutch and Swedish transport planners are beginning to adopt a staged approach that further brings together the two camps in a logical way. Thus, Voerknecht (2010) and Johansson (2010) propose (different, albeit similar) processes aimed at implementing improvements, ranging from behavioural change modification programmes to large scale additions to capacity.

Voerknecht suggests that sensible urban planning ought to be considered first, followed by more accurately pricing road space, mobility management (behavioural change) measures, public transport improvements, methods to enhance the efficiency of existing infrastructure, small scale modifications to the infrastructure to address pinch points in capacity, and only finally large scale additions to capacity. Meanwhile, Table 1.6 shows the steps suggested by Johansson and some of the implications that emerge from such a process.

Table 1.6 Implementing transport improvements in stages

Stage	Adopted measure	Approach	Impact	Implementers
1	Short-term behaviour change	Demand-side	Low cost, short-term, neighbourhood scale	Mobility managers
2	Long-term behaviour change			
3	More efficient use of infrastructure (soft measures)	Supply-side		Traffic and transport planners
4	More efficient use of infrastructure (hard measures)			
5	Small scale infrastructure improvements (public transport)			Transport engineers
6	Small scale infrastructure improvements (private transport)			
7	Large scale infrastructure improvements (public transport)			
8	Large scale infrastructure improvements (private transport)		High cost, long-term, national/international scale	

Source: Adapted from Johansson (2010)

In particular, it is interesting to note how the demand-side options tend to be the inexpensive, short-term and localised schemes, while the supply-side schemes are much larger and more time consuming and expensive, and also to note how the implementers change as one progresses through the stages. Communication across the involved organisations (and departments within organisations), then, would seem to be crucial in making this approach work.

However, while it is true that active city-level administrations can make a dramatic difference to the performance of the transport systems and hence to the quality of life in their localities, it is also the case that the next logical extension to the Package Approach – that is its application at a more localised scale as the 'Travel Plan' – remains limited.

Chapter 2 will therefore introduce the Travel Plan concept and explain its development so far – including the possible barriers to its adoption.

Traffic congestion in Cairo: negative externalities such as congestion are set to grow significantly, particularly in less developed countries

Gothenburg was one of the first major cities to adopt a comprehensive transport package of measures in the 1970s

References

Ball, P. 2004. *Critical Mass: How One thing Leads to Another*, (London: William Heinemann).

Bissell, D., Adey, P. and Laurier, E. 2011. Introduction to the special issue on geographies of the passenger, *Journal of Transport Geography*, 19(5), 1007–1009.

Cole, S. 2005. *Applied Transport Economics: Policy, Management and Decision Making*, (London: Kogan Page).

Dennis, K. and Urry, J. 2009. *After the Car*, (Cambridge: Polity Press).

Department of Transport (Great Britain). 1988a. *Roads for Prosperity*, (London: HMSO).

Department of Transport (Great Britain). 1988b. *National Road Traffic Forecasts*, (London: HMSO).

Department for Transport (Great Britain). 2010. *Transport Trends 2009*, (London: Department for Transport).

Department of the Environment, Transport and the Regions (Great Britain). 1998. *A New Deal for Transport: Better for Everyone*, (London: The Stationery Office).

Faulks, R. W. 1977. *Principles of Transport*, (London: Ian Allen).

Goodwin, P., Hallett, S., Kenny, F. and Stokes, G. 1991. *Transport: The New Realism*, (London: Rees Jeffreys Road Fund).

Headicar, P. 2009. *Transport Policy and Planning in Great Britain*, (London: Routledge).

H. M. Treasury and Infrastructure UK (Great Britain). 2010. *National Infrastructure Plan 2010*, (London: Crown Copyright). [Online] Available at http://www.hm-treasury.gov.uk/d/nationalinfrastructureplan251010.pdf [accessed 24 March 2011].

Hoyle, B. and Knowles, R. 1998. Transport geography: an introduction, in *Modern Transport Geography*, edited by B. Hoyle and R. Knowles, (Chichester: Wiley), 1–12.

Institution for Highways and Transportation. 1997. *Transport and the Urban Environment*, (London: Institution for Highways and Transportation).

Johansson, J. 2010. ResSmart – the platform for sustainable travel and transport in the Stockholm region. Paper presented at Moving People Bridging Spaces, *European Conference on Mobility Management*, Session 18, Graz, Austria, 5–7 May.

Knowles, R., Shaw, J. and Docherty, I. 2008. *Transport Geographies: Mobilities, Flows and Spaces*, (Oxford: Blackwell).

May, A. D. 1986. Traffic restraint: a review of the alternatives, *Transportation Research-A*, 20(2), 109–121.

May, A. D. 1991. Integrated transport strategies: a new approach to urban transport policy formulation in the UK, *Transport Reviews*, 11(3), 223–247.

May, A. D. 1997. Transport policy, in *Transport Planning and Traffic Engineering*, edited by C. A. O'Flaherty, (London: Arnold), 42–79.

May, A. D., Kato, H., Miyamoto, K., Okazaki, M., Sperling, D. and Vichiensan, V. 2004. Transport and land use instruments for a better environment, in *Urban Transport and the Environment: An International Perspective*, edited by H. Nakamura, Y. Hayashi and A. D. May, (World Conference on Transport Research Society and Institute for Transport Policy Studies, Amsterdam: Elsevier), 99–190.

Meyer, M. D. 1999. Demand management as an element of transportation policy: using carrots and sticks to influence travel behaviour, *Transportation Research Part A*, 33, 575–599.

Organisation for Economic Co-operation and Development. 1994. *Congestion Control and Demand Management,* Road Transport Research, (Paris: OECD).

Organisation for Economic Co-operation and Development and International Transport Forum, 2010. *Trends in the Transport Sector 1970–2008*, (Paris: OECD).

Ortuzar, J. de D. and Willumsen, L. G. 1994. *Modelling Transport*, (Chichester: John Wiley).

Parkhurst, G. and Dudley, G. 2008. Roads and traffic: from 'predict and provide' to 'making the best use of', in *Traffic Jam: Ten Years of 'Sustainable' Transport in the UK*, edited by I. Docherty and J. Shaw, (Bristol: The Policy Press), 51–73.

Schiller, P. L. Bruun, E. C. and Kenworthy, J. R. 2010. *An Introduction to Sustainable Transportation: Policy, Planning and Implementation*, (London: Earthscan).

Schreyer, C., Schneider, C., Maibach, M., Rothengatter, W., Doll, C. and Schmedding, D. 2004. *External Costs of Transport Update Study*, Final Report, (Paris: International Union of Railways).

Shaw, J. and Docherty, I. 2008. New deal or no deal? A decade of sustainable transport in the UK, in *Traffic Jam: Ten Years of 'Sustainable' Transport in the UK*, edited by I. Docherty and J. Shaw, (Bristol: The Policy Press), 3–28.

Shaw, J., Knowles, R. and Docherty, I. 2008. Introducing transport geographies, in *Transport Geographies: Mobilities, Flows and Spaces*, edited by R. Knowles, J. Shaw and I. Docherty, (Oxford: Blackwell).

Sperling, D. and Gordon, D. 2009. *Two Billion Cars: Driving Toward Sustainability*, (Oxford: Oxford University Press).

Standing Advisory Committee on Trunk Road Assessment. 1994. *Trunk Roads and the Generation of Traffic*, Report of the Standing Advisory Committee on Trunk Road Assessment, (London:The Stationery Office).

Steiner, R. L. 1992. Lessons for transportation demand management from utility industry demand-side management, *Transportation Research Record,* 1346, 14–17.

Voerknecht, H. 2010. Mobility scan: a revolutionary software tool to include MM in urban planning. Paper presented at Moving People Bridging Spaces, *European Conference on Mobility Management*, Session 18, Graz, Austria, 5–7 May.

Vuchic, V. R. 2007. *Urban Transit: Systems and Technology*, (Hoboken: John Wiley and Sons).

Warren, J. P. and Enoch, M. P. 2010. Island transport, car ownership and use: a focus on practices in Cuba, Malta, Mauritius and Singapore, *Island Studies Journal*, 5(2), 193–216.

Willoughby, C. 2001. Singapore's motorisation policies 1960–2000. *Transport Policy*, 8, 125–139.

Wright, P. H. and Ashford, N. J. 1998. *Transportation Engineering*, (Chichester: Wiley).

Chapter 2

The Development of the Travel Plan

1 What is a Travel Plan?

Emerging from the previous chapter was the idea that the concept of the 'Travel Plan' might be worthy of further study. This chapter first defines what this concept might be before providing a brief history of their evolution so far and then dismantling the concept into various building blocks to look at them more closely. It concludes by providing a map which lays out four key ways in which Travel Plans have already developed (and may in future develop) which in turn form the basis of subsequent chapters.

1.1 Agreeing Terms

For the purposes of this book, the term 'Travel Plan' covers a myriad of alternative descriptors. These range from 'site-based mobility management', 'site-based transport demand management', 'workplace transport demand management', 'employer-based trip reduction programmes', 'company mobility management', 'corporate mobility management', 'employer transport plans', 'employer commute option programmes' and 'green transport', 'green commuter' or 'mobility' plans and are illustrated in Table 2.1.

In general, variants of travel, transport or transportation 'demand management' (TDM) seem to be favoured in the United States, 'mobility management' (MM) in mainland Europe, and 'transport plans' or 'Travel Plans' in the UK. Moreover, to some extent the dominant motivations behind the Travel Plan concept in each region are then further revealed in the (sometimes changed) precursors.

Thus, 'commuter', 'corporate', 'company', 'worksite', 'school', 'employer' and 'site-based' suggest an emphasis on the type of agent delivering the instrument and/or the 'target market' of the measure while 'commute trip reduction' highlights a focus on transport objectives and the term 'green' indicates the importance of the environmental agenda as a wider societal objective.

Table 2.1 Defining the Travel Plan

Instrument	Provenance	Precursors		
		Delivery agent or target-market	**Objective related (transport emphasis)**	**Objective related (wider emphasis)**
Transport or transportation demand management	North America	Employer (employer-based), site-based, worksite	Commute trip reduction, trip reduction	
Mobility management	Europe	Company, commuter, corporate, site-based		
Plan	UK	Commuter, employer, school, station, residential	Transport, travel	Green

Ironically the term 'Travel Plan' is probably the most vague (and arguably the least suitable of all) but is deemed to suffice for now – it being the currently recognised UK term.

2 Defining the Travel Plan

Compared with agreeing the terminology, fixing definitions in this case is relatively straightforward. In examining this, the three 'provenances' from which Travel Plans have emerged will be looked at in turn and overall emerging themes reported.

2.1 Site-based Transportation Demand Management

Looking at the North American literature there are a number of definitions of general TDM. Typical of the genre are: "Travel Demand Management is a term used to describe a wide variety of techniques geared toward travel behaviour modification" (Ferguson 1997: 353); and "TDM can be defined as specific strategies to influence travel behaviour by mode, time of day and day of week, frequency, trip length, route, cost or regulation. This is to aid in fostering increased efficiency of a transportation facility or system" (Hendricks 2008: 30).

Meanwhile Berman et al (2006: 3) states that traditionally in the United States, TDM sought "to induce commuters to shift to higher occupancy modes, such as carpooling, vanpooling,[1] and public transport", but that "this view is no longer consistent" with what was happening currently. Instead, it notes that "a contemporary understanding of demand-side strategies is broader in scope than prior, more traditional views of TDM" and is now "about providing travellers, regardless of whether they drive alone, with travel choices, such as work location, route, time of travel and mode. In its broadest sense, demand management is defined as providing travellers with effective choices to improve travel reliability". Next, MTPWWM (1992: 5) sees TDM being something more integral to life in general – "TDM is a concept which you apply – automatically – in your daily life, for example travel allowances, company cars and goods transport. But you didn't know that it had a name".

Finally, Flynn and Glazer (1989: 11) makes the distinction between 'actions' and 'strategies'. TDM actions are measures that directly affect commuters, while TDM strategies refer to "a higher level government policy or programme consisting of ways to encourage or require intermediaries – such as employers, developers, transportation management organisations or governmental entities – to carry out [those] actions".

Focusing on the organisational or site-based level, Zuehlke and Guensler (2007: 171) suggests "employer-based trip reduction strategies urge employers to implement incentives to reduce single occupancy vehicles rates, emission rates and congestion by influencing employee commute behaviour".

Summing up, TDM is an alternative and complementary approach to the traditional approach of increasing the capacity of the transport system for national, regional and local government agencies in order to increase the efficiency and effectiveness of the transport network to solve transport problems like traffic congestion by modifying the behaviour of transport users. Consequently its main emphasis is on 'hard' policy mechanisms such as road space, parking and public transport management (including pricing and rationing), while so-called 'soft' measures like public transport marketing campaigns, encouraging people to share cars and persuading people to walk more are rather less visible.

Within this, one strategy – site-based TDM – typically relies on organisations buying in 'off-the-shelf' measures from a local or regional authority, transport operators and/or a TDM service provider. This would be done either to solve a specific transport-related problem, or to take advantage of a particular benefit (particularly public transport commuting-related tax incentives) or else in a few areas to meet the requirements of a Commute Trip Reduction programme or environmental regulation. In most cases, the objectives of the exercise are very much oriented to the goals of the organisation, and once the measure(s) (often only one or two rather than a package) is/are introduced then the role of the organisation is minimal.

1 Vanpooling refers to vans provided by an employer to an employee, who then collects and drives a group of colleagues between their homes and the workplace.

2.2 Site-based Mobility Management

"Mobility Management can be seen as a complementary, cost-effective approach to help raise the quality of mobility-related services. It is an innovative demand-oriented approach that establishes new partnerships to provide quality mobility services" (MOST 2008: 1). Expanding this, MAX Consortium (2007: 9) states:

> Mobility Management (MM) is a concept to promote sustainable transport and manage the demand for car use by changing travellers' attitudes and behaviour. At the core of MM are 'soft' measures like information and communication, organising services and coordinating activities of different partners. 'Soft' measures most often enhance the effectiveness of 'hard' measures within urban transport (for example, new tram lines, new roads and new bike lanes). MM measures (in comparison to 'hard' measures) do not necessarily require large financial investments and may have a high benefit-cost ratio.

In outlining the difference between TDM and MM, Wilhelm and Posch (2003: 174) suggests that whereas TDM focuses mainly on the journey to work and on the formation of carpools and vanpools, MM is "a much broader concept for all modes of transport and all aspects of mobility". Building on this idea, MAX Consortium (2007: 9) notes that:

> in many countries, MM is predominantly a site-based activity connected to a traffic generating site such as a company, a school, concerts, sports matches, fairs, hospitals, entire administrations based in a number of locations, recreational sites and so on. In these cases MM seeks to manage the way in which people travel to the site in question.

Interestingly, a different perspective is offered in Berman et al (2006: 4), which instead of market niches considers measures and sees MM as being more narrowly focused on measures such as information, marketing, partnerships, communications, and promotion of sustainable modes. "In a sense, [in terms of measures] MM might be viewed as a subset of measures within the broader U.S. definition of TDM".

Once more looking at the organisational level, "company (commuter) mobility management (CMM) is about: improving mobility effectively through employers introducing measures to influence the business related movement of their staff/ employees (and of goods) to reduce avoidable solo-use of the car" (Novem undated: 1). Or, in other words, a company mobility plan is a strategy that includes a series of measures such as public transport promotion, company buses, car-pooling, cycling, walking and introduction of measures such as parking management to better use the organisation's resources by combining transport and other business issues in a co-ordinated package (Traject et al 2000).

Whilst the above is sufficient as definition, working more in depth with MM can pose some questions, specifically relating to its scope (MAX Consortium 2007: 9–10). Thus:

- MM is demand-oriented – instead of supply-oriented. This means that building new tram lines, new bicycle paths, new roads and so on are not considered to be MM, as these are all supply-side measures. This is reflected, for example, in the policies of the Netherlands and Sweden, where it is in many cases obligatory to first look at the potential of MM before permission for supply – side measures like road building is given. It is also reflected in the fact that in the USA and some other countries MM is called Travel Demand Management (TDM).
- Infrastructure measures can be supportive measures for MM. In many countries, MM is seen most often as a site-based activity – connected to a traffic generating site such as a company, a school, or a tourist attraction. On sites, such as hospitals or business parks, a negotiated MM measure package might include infrastructure (bicycle parking, tram stops, car parking, a bus shuttle service – typical supply measures). They are considered as supportive measures (but not as MM).
- MM does not necessarily have to be limited to a site. City-wide, regional or even national appliances like car-sharing (car clubs in the UK), carpooling, mobility centres, new resident information packages, campaigns, integration of several services into one fare system are regarded as MM.
- Sustainable urban transport plans are not MM, but they should contain MM. MM does not include the whole spectrum of traffic and transport planning. Transport plans are part of MM when they are site-based, such as Workplace Travel Plans or School Travel Plans. The official term coined for this by Momentum is the mobility plan. HOV-lanes, congestion charging, parking management and road tolling, whilst being typical demand-oriented measures, are not MM, but can be supportive measures for MM. As parking management is often central to site-based MM, parking management as part of a MM bundle of measures is considered to be an integral part of MM.
- Traffic system management is not considered part of Mobility Management. However, components of traffic management that aim to influence demand and change attitudes, and especially if they make the use of alternatives to the car more attractive, be it by delivering real departure time of trains on mobile phone or internet or variable message signs or via navigation systems, are considered as MM.
- Raising travel awareness, mobility education, and the marketing of sustainable modes are each regarded as part of MM. Some MM theoreticians regard these as being quite distinct. In practice, it is almost impossible to draw any boundaries.
- MM is considered to encompass goods transport, as long as it is site-based and the measures concerning goods are part of a mobility plan that also includes

passengers. In any case it includes the transport of passenger luggage. But if these conditions do not apply, organising goods transport is considered as logistics for which there already exists a highly specialised industry.
- Finally, various legislation, pricing incentives and disincentives are part of MM, if they support concrete MM measures that fall within the demarcation as described above.

Adding to this, Attenbrand et al (2005) reports that mobility management has the following attributes:

- Several actions/services to provide alternatives to car use.
- Actors involved should be those who generate traffic.
- Actions are voluntary to end users.
- There is also a need for supporting measures from public administrations.

At the conceptual level then, Mobility Management can be seen as being an umbrella term for a series of softer transport-related measures. It is therefore unlike TDM as it does not include the 'harder' policy measures like road user charging, removing parking spaces or subsidising bus fares, possibly because such measures are already far more prevalent in Europe as a traditional part of the transport planner's toolbox than in North America. Meanwhile the motivations for action stem almost exclusively from public authority policy concerns. Accordingly, the objectives are much more oriented to public policy goals that are broader than transport than for TDM. These include mitigating climate change, limiting energy use and improving local air quality (environment), and improving access to facilities by means other than the car and people's health (social), as well as reducing delays and travel time uncertainty caused by congestion (economic).

With the exception of experiences in the Netherlands and (several years ago) Italy, site-based Mobility Management so far seems to have been limited to a number of isolated 'pilot' localities spread around Europe (for example Ghent, Zurich, Paris, Dublin, Vienna).

Traditionally, most plans have been imposed by national, regional and/or local government through environmental regulation requiring organisations over a particular size to employ a Travel Plan coordinator or to mitigate their transport impacts, and relatively few have been introduced voluntarily. This, though, seems to be changing with several 'progressive' authorities now shifting their efforts towards encouraging organisations through improved information support and financial inducements.

Looking at plan objectives, as with the motivations these are necessarily fully aligned with organisational interests, and are more often related to the aims of the wider community (often to improve air quality or reduce congestion). Site-based mobility plans are seen as being successful if the organisation is actively involved, though the idea of the 'package of measures' seems fairly low-key in general.

2.3 Travel Plans

With the UK model, Travel Plans are currently incorporated under the 'Smarter Choices' moniker, which was coined in a report made for the Department for Transport to cover ten 'soft' (effectively mobility management) measures, namely: Workplace Travel Planning, School Travel Plans, Personalised Travel Planning, public transport information and marketing, travel awareness campaigns, car sharing (that is carpooling or lift sharing), car clubs (the UK term for car sharing), teleworking, teleconferencing and home shopping (Cairns et al 2005). This new term – which refers to "techniques for influencing travel behaviour towards more sustainable options, such as walking, cycling, travelling by public transport and car sharing" (Cairns et al 2005: 3) – was devised to avoid confusion. This is because in the UK context the term mobility management evokes thoughts of either enhancing accessibility for people who are mobility impaired on the one hand, or about how to improve the employment prospects of socially excluded groups on the other.

Looking at the UK literature on Travel Plans then, a 'traditional' definition is:

> a general term for a package of measures tailored to meet the needs of individual sites and aimed at promoting greener, cleaner travel choices and reducing reliance on the car. It involves the development of a set of mechanisms, initiatives and targets that together can enable an organisation to reduce the impact of travel and transport on the environment, whilst also bringing a number of other benefits to the organisation as an employer and to staff (EEBPP 2001: 1).

Similar definitions are offered in Rye (2002: 180), which notes that a Travel Plan "provides a strategy for an organisation to reduce its transportation impacts and to influence the travel behaviour of its employees, suppliers, visitors and customers", and in DfT and NBTN (2008). However, more recent understandings have begun to change slightly to put more emphasis on Travel Plans as a business tool instead of as a transport policy tool, such that BSI (2008: iv) sees a Travel Plan as "a long-term management strategy for an organisation and its various sites or business park that seeks to deliver transport objectives through positive action and is articulated by a document that is regularly reviewed" – see Chapter 5 for more on this.

In summary, under the UK model, Travel Plans emerged as one of several alternatives to supply-side enhancements once the effectiveness of road building as a congestion reduction policy was undermined during the early mid-1990s. Since named as Smarter Choice measures, these roughly correspond to Mobility Management measures in scope, but for reasons of language had to be labelled differently.

However, while MM measures may correspond to Smarter Choice measures on one level, at the site-based level there are differences as well as similarities. Thus, unlike in Europe, Travel Plans in the UK started out largely as being delivered voluntarily by organisations to address (transport-related) problems, and it was only a little later that local authority regulations (through the planning system) began to mandate their introduction so as to meet the needs of the wider community. Travel

Plans too are seen as needing a comprehensive package of (transport-focused) measures that fully take account of the site characteristics. There has thus been less scope for 'off-the-shelf' measures than in the US context and hence organisations are possibly intimidated by the potential commitment involved. As in Europe, but unlike in North America, the active involvement of organisations in managing transport is seen as being positive.

2.4 Emerging Themes

From the above definitions, the key features of the site-based TDM, site-based MM and Travel Plans are summarised in Table 2.2.

Table 2.2 Site-based TDM, MM and Travel Plans compared

	Site-based Transportation Demand Management	Site-based Mobility Management	Travel Plan
Delivery mechanism	Site-based TDM strategies an umbrella term for toolbox of measures.	Outside of the Netherlands and Italy site-based TDM still restricted to a fairly small number of localities.	Travel Plans very much a delivery mechanism of other measures.
Additional agent	Role of organisation crucial but often minimal once measure(s) in place.	Role of organisation looks like becoming more important.	Ideal is where organisations are actively involved.
Motivations	Organisational self-interest now dominant. Some areas where public authorities require Travel Plans but mostly incentive rather than penalty based.	'Sold' to organisations by public authorities as a means of meeting societal obligations, often supported by regulatory sticks. Organisational benefits generally downplayed.	Two categories. Obtaining planning permission a major driver, whereby the local authority seeks to mitigate impacts of new development on society. Also, organisations looking to address transport-related problems (poor site access, parking).
Objectives	Addressing transport problems (traffic congestion, poor site access, increasing transport costs). Air quality too sometimes.	Addressing congestion, air quality, energy reduction, climate change, public health and accessibility goals for society at large.	Mix of societal objectives (congestion, environmental and social) and organisational (site access, parking, CSR, environmental).

Site-specific	Only in special circumstances. Instead, measures/ tools available 'off-the-shelf' from TDM service providers, local transport operators and local authorities.	Site-specific measures only for large organisations. Most mobility management measures introduced on a regional basis by public authorities and transport operators.	Ideal Travel Plan measures are tailor made to suit the site context. Relatively few 'off-the-shelf' tools available from transport industry (with a few exceptions).
Package	Idea of a package of measures at site level fairly weak. Often individual measures applied by organisations instead.	Idea of a package of measures at site level fairly weak. Often individual measures applied by organisations instead.	Package of measures is a strong element in Travel Plans.

From Table 2.2, the common themes that emerge to underpin the Travel Plan concept, are:

- Travel Plans are not really an instrument themselves, but a delivery mechanism or strategy for other mostly transport-focused measures.
- Travel Plans are delivered by an additional 'agent' that is not a part of the 'traditional' transport policy institutional structure.
- Travel Plans are initiated in two ways: either by the organisation or by government.
- Travel Plans seek to deliver transport and related benefits to the community as well as some more direct organisational benefits to the participating 'delivery agents'.
- Travel Plans are to some extent 'site-specific', that is tailored to the specific contextual circumstances.
- Travel Plans to some extent deliver a package or a strategy of a 'wide variety' of transport instruments.

Overall, then, a catch all definition that emerges from these points is that Travel Plans are a mechanism for delivering a package of transport measures targeted at a specific site by an agent with a strong relationship with the local transport users to deliver transport and wider goals to the organisation and society as a whole.

Before these themes are further explored, however, it is useful to present here some context as to how this concept first emerged.

3 A Short History of Travel Plans

Travel Plans, as we now understand them, are widely ascribed by commentators to have first emerged in the 1970s in the United States (though several forms of closely related mechanisms were previously known as we shall see later in the chapter). This section looks first at their evolution in the USA, and then at their dissemination to the Netherlands, the UK, Italy and beyond. Please note that data on the current (2011) international status of Travel Plans, and comments on the potential future of the concept is provided towards the end of Chapter 7.

3.1 Travel Plans in the United States

Schreffler (1986) notes that Travel Plans first emerged as a result of the oil crises in 1973 and 1979, which prompted a 'vast number' of employers to adopt a range of strategies such as subsidising bus passes, flexi-hour schemes, providing works buses and vanpools and supporting carpooling. Martz (2006) and Winters (2010) provide examples such as Amoco in Houston, 3M in Minneapolis-St Paul and Aerospace Corporation in Los Angeles where such practices occurred.

Schreffler continues that what had been a private sector response then began to be recognised by the public sector, leading to the US Federal Government then authorising money for more than 100 TDM demonstration projects through the 1974 Emergency Highway Energy Conservation Act.[2] Government intervention through regulation then developed throughout the 1970s and 1980s through Federal requirements for transportation system management at the regional level in the form of Air Quality Management Districts in areas with poor air quality particularly under the 1977 Amendments to the 1970 Clean Air Act. In practice, several air quality agencies therefore adopted TDM regulations, the most well-known of which was Regulation XV of the Southern California Air Quality Management District adopted in December 1987. This required companies with more than 100 employees to employ a trained Employee Transport Coordinator to help meet targets set by a Trip Reduction Program to help reduce pollution levels (Winters 2010; Dill 1998). Probably the highpoint of this legislative push at the Federal level came in 1990 with the Employee Commute Options (ECO) requirement of the Clean Air Act Amendments, which mandated large employers in the most polluted cities to submit annual trip reduction plans.

Also in the 1980s, traffic congestion became a major concern to local authorities while many suburban areas also faced pressure from developers to allow the building of large new office, industrial or retail parks. This meant that by 1990 a number of cities required TDM measures as a condition of planning consent, while around 50 cities adopted so-called Trip Reduction Ordinances – whereby new and existing large employment complexes were required to reduce their trips (Dill 1998).

By the early 1990s however, such a heavy regulatory approach to Travel Plans was starting to create a backlash by business interests, such that in 1995

2 Interestingly, one key finding from this was that the ongoing involvement and commitment of employers was a key weakness, and a reason for this was because employers were required to invest significant resources in surveying employees with little say in the process. Consequently, Schreffler notes that 'third-party' and 'brokerage' agencies emerged to provide mobility services in the late 1970s.

Regulation XV was abandoned following a legal challenge, and the ECO became a voluntary programme. Accordingly, as of the present day only in Washington State is there a large scale mandatory Commute Trip Reduction Program still in operation, although numerous other 'voluntary' and smaller scale schemes remain. More recently therefore, local, regional and national government has shifted away from the use of regulations towards other support mechanisms such as information and fiscal incentives (see Chapter 6).

3.2 Travel Plans in the Netherlands

At this point, the story shifts to the Netherlands which Hoogma (2006) suggests can be broken down into four stages.

Firstly, in 1989 the Company or Commuter Mobility Management concept was transferred from the United States to the Netherlands following a study trip by a group of Dutch transport experts, and was adopted in the SVV II (the Second Transport Structure Plan White Paper) a year later. Under this, experiments were established in organisations across the country and a plan was adopted that envisaged all companies with more than 50 employees having a Travel Plan.

Secondly, from 1994 to 1996 the process accelerated due to: congestion worsening faster than expected; the introduction of more public money and tax breaks for company mobility management; and because a report from management consultants McKinsey suggested that mobility management techniques were 15 times cheaper than equivalent road construction.

Third, from 1997 to 2001 there was a process of decentralisation, whereby powers were devolved to regional and local governments. During the same period, policy goals for Travel Plans were broadened to the environment and energy efficiency sectors, while interventions became more focused in areas where there were recognised accessibility issues.

Fourth, between 2001 and 2006 at the national level there was relatively little activity beyond a few minor tax changes, while at the regional level the focus switched to looking at parking standards and regional promotion points.

Interestingly, since 2006 the Dutch Government has considered adopting a so-called 'points-based plan' whereby companies would be able to include mobility management measures towards meeting their environmental commitments. In the event, though, companies have pledged to voluntarily introduce measures to reduce car use through so-called 'covenants' (see Chapter 6). Moreover, Travel Plans have also become far more focused on the needs of organisations to the extent that (explicit) objectives related to reducing car use are no longer mentioned (Metz, F. 2011. Personal communication, Director, KPVV, Amsterdam, 24 February).

3.3 Travel Plans Elsewhere

Subsequently, the Travel Plan concept spread to countries including the UK, Italy, Australia and parts of Belgium and Germany.

In the UK, organisations such as Boots (a pharmaceutical company) and Nottinghamshire County Council voluntarily adopted Travel Plans in 1995, as did various hospitals such as Derriford in Plymouth and Addenbrooks in Cambridge. Travel Plans then spread to other sectors, notably schools, airports, leisure facilities, residential developments and individuals. National policy support was first provided in the White Paper *A New Deal for Transport: Better for Everyone* (DETR 1998), and most recently in *Towards a Sustainable Transport System* (DfT 2009). One other important point to note here is the increasing importance of mandatory Travel Plans through local planning authorities requiring developers to submit a Travel Plan as part of the process for obtaining planning permission – and this is currently the major 'source' of UK Travel Plans (to be discussed in more detail in Chapter 6).

In Italy too, the mandatory approach has been the main approach adopted to Travel Planning. Here, Bradshaw and Camara (1999) reports that a series of Decrees were approved from 1998 that aimed to tackle congestion and pollution. So, in November of that year, guidelines were issued to mayors on how to achieve pollution improvements, followed by approval for a five-year carbon tax the following month. Also in December, the almost 3000 private companies employing more than 300 people were charged with appointing a mobility manager to optimise commuting and to reduce car use, a proposal which was to be developed further in 1999 when mobility managers were also to be introduced in schools.

More recently, as of 2010, Travel Plans are finally being launched at a national level in countries as diverse as France, Denmark, Finland and Japan, and at a regional basis in Catalonia (Spain), and parts of Switzerland, apparently presaging the mainstreaming of the idea internationally. On the other hand, evidence from parts of Australia (Victoria for example) suggests that Travel Planning there is in retreat, while the global recession threatens to at least partially address many of the issues (such as congestion, shortage of space and parking difficulties) that Travel Plans are currently geared towards solving. The future therefore, would seem to be in the balance (see Chapter 8 for more on this).

4 Travel Plan Themes Explored

With this situation in mind, the next step is to look at each of the themes identified at the end of Section 4.2 in turn.

4.1 Travel Plans as a Delivery Mechanism for Other Measures

Running throughout the following 'sub-themes' is the understanding that Travel Plans, unlike other transport tools, are in fact a delivery mechanism and not an instrument in themselves. One significant implication of this relates to determining the most appropriate institutional level of delivery for applying different public authority roles for each available policy area or instrument. Thus, for transport the institutional levels of delivery range from International (global, continental) through National, Regional (sub-national, provincial, regional), Local (county, city, district) and Sub-local (community, ward, neighbourhood, parish, site). Public authority roles can be categorised as being one of three types:

- Macro – here, strategic decisions are taken about how the ideas and values of policy makers are transferred into practical outcomes. These might include establishing operating regimes, setting policy objectives; agreeing environmental, safety and competition standards; and fixing resource levels.
- Meso – planning decisions are made at the meso level to do with planning operations; licensing routes, vehicles and drivers; and with allocating resources to specific budgets.
- Micro – the focus at this scale is operational and about service delivery. It is about building, operating, maintaining and enforcing and about spending resources.

Transport policy areas in this sense could be applied on a 'modal' (for example car, bus, bicycle, walk) basis, by journey distance or perhaps by journey purpose. Policy instruments could include informational, regulatory, fiscal or technological measures (as described in Chapter 1). In combining the institutional level of delivery, the public authority roles and the policy areas, examples include how international maritime and aviation issues tend to be dealt with at the global and continental levels, whereas the most appropriate institutional level for even 'macro-level' decisions about infrastructure for walking may be at the sub-local level. Of course between these extremes there is considerable scope for different societies with different contextual circumstances to arrive at different answers as to what the most appropriate institutional delivery level is. Hence, in some countries (such as France) most functions are concentrated at the national level while in others (Canada and Germany for example) the provincial or regional level has a more prominent role.

There are two issues on which to make a judgement in this 'vertical integration' dimension. The first concerns the optimal geographical scale of each administrative unit to deliver transport and transport-related functions, and the second the optimal number of layers of administration needed to deliver transport and transport-related functions.

Table 2.3 illustrates the advantages and disadvantages between policy being applied centrally and at a more devolved level, while Table 2.4 summarises the tensions involved between multiple tier and single tier institutional arrangements.

Table 2.3 Centralised versus devolved delivery systems

	Centralised delivery policy applied through (fewer) larger centralised units	Devolved delivery policy applied through (more) smaller devolved units
Benefits	Good coordination. 'Joined-up' policy decisions. Economies of scale. Good skill set.	Focus. Responsive to local needs.
Problems	Unwieldy. Lack of focus. Less responsive to local needs.	Fragmentation. Overlapping responsibilities. Fiefdoms. Skill shortages. Cost.

From Table 2.3 it can be seen that centralised delivery systems theoretically lead to better coordination of effort and economies of scale while more devolved delivery systems are seen to have more focus and be more responsive to local needs. Meanwhile Table 2.4 shows that single tier arrangements offer simplicity, but that multiple tier systems are better able to deliver policy at more appropriate levels.

Table 2.4 Single versus multiple tier delivery systems

	Single tier policy delivery systems	Multiple tier policy delivery systems
Benefits	Simple structure. 'Joined-up' policy decisions. Economies of scale.	Services delivered at most appropriate level.
Problems	Some services not local enough, others not strategic enough. Skill shortages.	Overlapping responsibilities. Complexity. Cost.

The challenge for the transport policy decision maker, then, is to balance the advantages and disadvantages of each in a way most appropriate to the context at the time. Evidence that this is not a straightforward task is widely available, and is illustrated through continual re-organisations within both public and private sector authorities and organisations.

4.2 Delivery by an Additional 'Agent'

The creation of a new agency or 'intermediary' such as an employer, a traffic generating organisation more generally or some other form of organisation within the transport policy institutional structure would seem to be of significance, and in many ways makes a good deal of sense. This sentiment is perhaps best expressed in Comsis (2002: I.1–I.2), which states:

> Employer-based TDM programs have long been the focus of many public agency TDM efforts. This stems, in part, from the recognition that employer policies and procedures, such as free parking, overtime and so on, can be a significant factor in influencing how employees choose to get to and from work. Moreover, since employers have immediate and direct access to their employees, they provide a means by which public agencies can leverage relatively limited public resources for encouraging alternatives to solo driving … [By contrast] with the exception of their own employees, the capacity of public agencies to interact with the commuting public in this manner is severely limited.

This change in approach can be represented as in Figure 2.1.

Figure 2.1 shows that, under the traditional transport planning approach (see left hand side), local authorities seek to communicate directly with individuals in the community. However, this relationship (if there is one at all) is generally weak, meaning that the chance of a local authority influencing how any individual might behave is small. To address this lack of influence, the Travel Plan concept (see right hand side of diagram) sees local authorities first contacting employers (step 1) who in turn then pass on the 'Travel Plan message' to their employees (step 2). This latter communication channel can potentially be very effective, because employers clearly can exert a significant influence over how their employees behave.

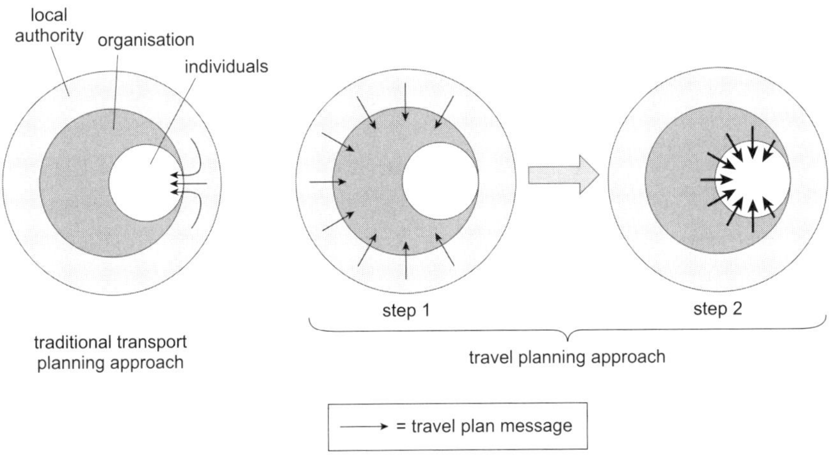

Figure 2.1 From local authority to organisation as delivery agent

Arising from this idea, Zupan (1992) specifically considers the TDM issue to present a range of TDM strategies and (perhaps surprisingly) concludes that the employer is most often the most suitable delivery agent (see Table 2.5).

Table 2.5 TDM strategies and suitable delivery agents

TDM strategies	Suitable delivery agents
Alternative work schedules (staggered, flexible, four-day, telecommuting)	Employer, TMA
Carpools	
Informal	Individual
Formal	Employer, TMA, county or state
Vanpools	As above, van brokerage companies
Subscription bus	Employer, TMA, county, transit operator
Parking management	
Preferential parking	Employer
Parking pricing	Employer
Parking ratios	County or municipality, developer
Park-and-ride	County or tate, developer
Preferential road treatments	Government, road authority, municipality
Congestion pricing	Government, toll road authority
Transit	
Transitchek	Transit agency, operator, employer
Employer subsidised transit	Employer, operator, TMA
Employer sponsored transit	Employer, operator, TMA
Transit coordinator	Employer, operator, TMA
Land use and zoning	
Higher densities	Municipality
Transit-friendly design	Municipality, government
Mixed use development	Municipality, county
Growth management	County and state
Trip reduction ordinances	Municipality
Transportation management associations	Employers, state, county

Source: Zupan (1992)

Explaining why this may be the case, Miller et al (2005) suggests that three types of organisations commonly deliver transport services: public agencies, private (for profit) and private (not for profit) organisations, and provides pros and cons for each (from a US perspective) in Table 2.6.

Table 2.6 Public and private sector delivery agents

	Public agencies	Private (for profit)	Private (not for profit)
Advantages	Can tax and exercise right of eminent domain to acquire property. Some grant programmes only available to public sector. Wider tax exemptions than private agencies.	Enjoys limited liability, some tax advantages and can receive public and private grants and donations. Can own assets, employ personnel and enjoy other benefits. More likely to be offered donations of time, money and other resources than alternatives so low costs. Generally likely to have similar culture to other 'client' organisations.	Enjoys limited liability, some tax advantages and can receive public and private grants and donations. Can own assets, employ personnel and enjoy other benefits. Immune from political barriers preventing service expansion.
Disadvantages	Political barriers preventing service expansion.	Cannot tax or exercise right of eminent domain to acquire property.	Often difficult to generate sufficient revenues to support transport organisations.

Meanwhile Darbera (2002) comes up with three core transport-related issues where he sees employers as being best placed to intervene (from a pragmatic rather than an idealistic perspective) (see Table 2.7).

Table 2.7 Reasons for organisational intervention in transport

Transport-related issue	Justification for employer intervention
Most journey to work trips made in rush hour, hence external costs to society imposed through congestion.	Complex issue and should be responsibility of government. However, government reluctant to introduce road pricing. Firms can subsidise non-drivers and either restrict or levy additional charges on private car use.
Vehicles immobile for most of the day consuming space which has a value but which the employer is providing free. Non-driving employees and customers therefore subsidising these car users.	Charge market price for spaces.
Decision to drive often due to imperfect information regarding alternatives.	May be straight forward for company to rectify.

Source: Darbera (2002)

As already reported, Travel Plans in their current form have been in operation since the 1970s, yet this tendency for employers to become actively involved in delivering 'additional' services to individuals stretches back much further in time. Thus, in transport, Dill (1998) reports that companies during the Second World War were required to adopt TDM policies aimed at conserving resources (rubber, steel, oil). Still earlier, from the 1920s many firms provided works buses to ensure their staff reached work on time. For example, Readers Digest relocated its headquarters from New York City to Westchester County and provided works buses from the new site to surrounding communities in order to ensure that staff could get to the office (Shreffler 1986).

Perhaps rather more influential in terms of employer support for individuals as a response to the commuting issue, is in providing housing near the workplace. Thus, Lund (2006) notes that in Victorian times in the UK employers provided 10 per cent of the housing stock, some of which was built in modern high quality villages. Similarly, looking at the US coal industry in the nineteenth and twentieth centuries, Fishback (1992) notes that in the typical company town the employer owned the store, the housing and often provided governmental services like sanitation and law enforcement. In fact, one might argue that landowners in Feudal agricultural systems (which extended across significant parts of Europe from the Middle Ages until up to the late nineteenth century) effectively provided housing for their 'employees'.

Returning to the present, though, the role of the employer today *vis a vis* the commute has largely appeared to have diminished when compared with historical levels. The notable exceptions here are of housing suppliers such as national Armed Forces and hospital authorities and organisations that still provide free parking spaces for staff (as well as the far smaller number of companies that have so far adopted Travel Plans).

Overall this process can be represented as in Figure 2.2.

Interestingly, the historical precedence point was used by the Dutch Government in a 1992 leaflet aimed at employers which states:

> seen from a historical perspective, employers have influenced commuting since the industrial revolution. The construction of houses for labourers near the factory is a prime example. Travelling hours were kept to a minimum and working hours were long. Not so many years ago in the [nineteen] fifties and sixties, many firms organised company transport to ensure that staff got to work on time. More recently, most companies restrict themselves to offering travel allowances, providing parking spaces and lease cars. But if we look at the current situation, the transportation demand management of the seventies and eighties is rapidly becoming outdated (MTPWWM 1992: 6).

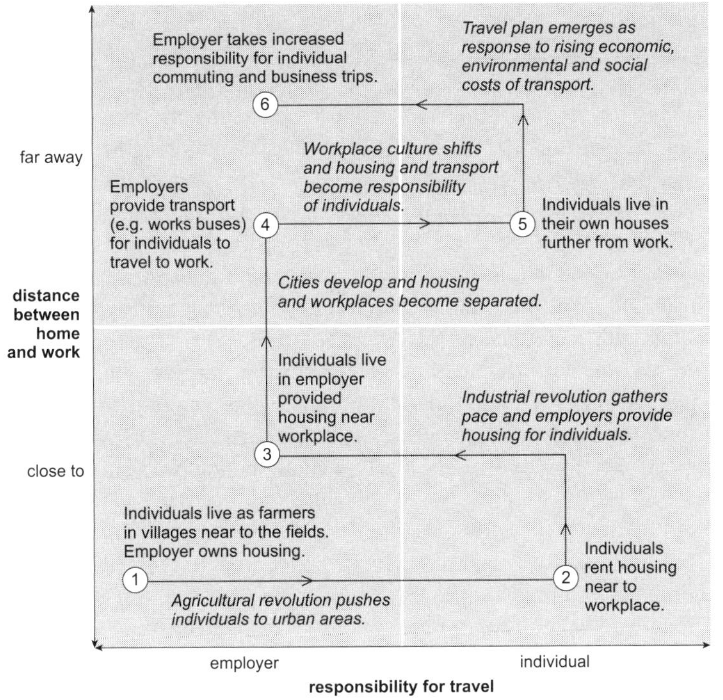

Figure 2.2 Changing approaches to employer and individual responsibility

To conclude, in the words of Attenbrand et al (2005: 13):

> when striving to reduce private car dependence many different and optional ways of travelling are needed. The Mobility Management concept points out that the private sector (employers, organisers of special events and so on) could and should carry its responsibility for offering a wide variety of travel modes: favouring those that have the least harmful impacts on the environment and that possibly have positive effects on public health. Reducing private car dependence is not the sole responsibility of public administrations.

4.3 Motivations behind Travel Plans

Examining the motivations behind why Travel Plans are established is clearly important if any understanding is to be gained as to what has happened (and is happening) in the sector. Simplistically, it could be argued that there are two core stakeholders with an interest in the introduction and operation of Travel Plans – public authorities and organisations.

Government motivations for establishing Travel Plans would seem to be fairly straightforward. Citing the Dutch context, MTPWWM (1992) sets out that providing new transport infrastructure will not keep pace with increase in traffic volume and suggests that there is a need to work closely with employers and employees to find an answer. To do this it emphasises focusing on commuters and on the peak hour congestion problem.

Such observations suggest that involving the employer as a mediating institution to deliver Travel Plans is a sensible idea – from the government perspective at least. However, while this is all very well in theory, it does rather assume that the employer will sign up to the same objectives as those set by the public authority. Unfortunately, this is not necessarily the case.

Instead, employer motivations are probably more complex and not necessarily complementary to those of the transport policy maker – a point which is often not sufficiently recognised by those policy makers. Thus, DeHart Davis and Guensler (2005) reports that employers voluntarily mediate policy problems either because of:

- Organisational self-interest (that is when there is perceived to be a direct or indirect economic payoff);
- Organisational control (where actions taken reduce employee discretion so that employees behaviour becomes more closely aligned to the goals of the organisation);
- Association membership (where peer pressure generates a socialisation process whereby identity is developed, norms are communicated and members then conform to these norms so as to gain the approval and respect of fellow members).

More directly Rye (2002) reports that the main drivers are estate management, accessibility and amenity; external regulation; image; and leading by example, and MTPWWM (1992) notes Travel Plan benefits to organisations as including increased accessibility, lower parking costs, site expansion potential, reduced travel allowances, less staff arrive late, improved supply and delivery, easier recruitment, positive contribution to company image, environmentally friendly, employees appreciate concern better local contacts.

In summarising these organisational motivations, it would seem that two distinct types of motivation emerge – specifically, that Travel Plans can either be adopted voluntarily by an organisation or else they can be imposed most usually by a local, regional or national government body. Beneath this the motivations can be further divided into economic, social and environmental. It is also interesting to note here whether such motivations are transport-led or not (see Table 2.8).

Rye (1997: 43) sums up the position perfectly:

> employer-based TDM is in the interesting position of being something which an organisation may attempt to implement but which, usually, is unrelated to the raison d'etre of that organisation … This situation is bound to result in conflict between those who are trying to implement the TDM programme and those who see it as a distraction from the firm's main activities. The organisational peripheral nature of TDM is not addressed in the existing literature, yet in the author's view, it is a – perhaps the – crucial issue which affects the success of TDM at any workplace.

To this Zupan (1992: 8) adds: "in a democratic society, the idea of artificially affecting individual choices for the greater good is not an easy concept to put in place. And without a clear and present danger as motivation, TDM may become just another transportation acronym".

Table 2.8 Travel Plan motivations

Voluntary	Economic	Social	Environmental
Transport-related	Reduce congestion, improve journey time reliability, save money and resources spent on parking.	Improve access for staff and relationships with neighbours.	Reduce local noise and air pollution.
Non-transport-related	Improved efficiency through improved staff productivity.	Corporate social responsibility, image.	Cut CO_2 emissions.
Mandatory	Meet legal obligations and/ or secure planning permission for new developments.		Meet CO_2 reduction obligations.

Looking at this issue from a different perspective, Herzog and Grant (2002) identifies the various benefits available to each of the different stakeholder groups (see Table 2.9).

Table 2.9 Possible employer, employee and community benefits

Employer	Employee	Community
Reduced costs: Financial incentive costs Facility costs Administrative costs	Direct financial benefits (including tax savings) Reduced driving expenses Employee parking cost savings	Reduced vehicle travel Reduced urban air pollution Reduced carbon dioxide emissions
Savings: Parking cost savings Office space cost savings		
Business benefits: Improved recruitment /retention Increased employee productivity		

Source: Herzog and Grant (2002)

From this, it is clear that the needs of each group are significantly different and that this is something that policy makers aiming to persuade organisations to take up Travel Plans need to recognise.

4.4 Objectives of Travel Plans

Strongly related to the motivations behind a Travel Plan are the objectives of what it is trying to achieve. Hence, Table 2.8 (see earlier) also implies that the plans enacted due to voluntary motivations tend to focus more on narrower business-focused outcomes, while the mandatory plans are often charged with meeting wider community-benefiting targets.

Fundamentally, then, at the conceptual level there are significant tensions between the idea of the Travel Plan as predominantly a business tool (generated by pressures within an organisation) and as predominantly a public policy tool (generated by externally applied pressure). This has several important spin offs, which will be explored further in Chapters 5 and 6:

- The first concerns the role of the regional or local public authority (is it/should it be a supporter, regulator or an uninvolved party?).
- Second, is the role of the Travel Plan organisation (is it/should it be active or passive, reactive or proactive?).
- Third, how should they relate to each other?
- Finally, how should organisations seek to incorporate Travel Plans into their internal operational structure?

4.5 The 'Site-specific' Nature of Travel Plans

As Thaler et al (2002: 14) points out, the "company management of staff's travel choices involves the formulation and implementation of company-specific approaches to transport with the object of achieving environmentally friendly, sustainable staff transport and rationalisation of transport". From this, the interesting dimensions relating to the site-specific nature of the location first centre on the optimal size of the Travel Plan organisation. Once again the issues here are framed with the objective of trying to determine the most appropriate level for Travel Plans to deliver suitable transport instruments and services. Regarding the issue of size, in theory, the tensions would be between the smaller and more focused but less well resourced plans on the one hand and the larger more efficient but less targeted on the other. Table 2.10 sets these out.

Table 2.10 Tensions as to the optimum size of a Travel Plan

	Smaller	Larger
Advantages	Plan can be more focused on the needs of the user.	Economies of scale allowing more efficient use of resources and a wider range of instruments to be introduced, and greater opportunities for partnerships with external agencies.
Disadvantages	Diseconomies of scale.	Reduced focus on individual user needs.

In practice, in the vast majority of cases, Travel Plan organisations strive to be as large as possible to the extent that those at the smaller end of the spectrum often seek to link up and merge with larger Travel Plans in their area in the form of groups or networks (an aspect which will be further explored in Chapter 4).

In terms of organisation type from a transport perspective, crucial aspects link to the trip types generated and their particular characteristics. Thus, for example, traditional office development generates regular peak flows in the morning and evening from Monday to Friday, with some less frequent business trips and delivery trips being made during the course of each day. By contrast, a football stadium may only host 25 or so events a year, but each may have the potential to paralyse the surrounding transport network for a significant period before and after the match (Chapter 3 will explore these issues further).

Also important are factors to do with the level of transport supply and its relationship with the demand (for example level of parking availability or public transport access). Here, organisations are only likely to be interested in addressing transport issues if they feel there to be a problem that impacts on their organisational objectives in some way, perhaps because of a shortage of parking spaces resulting in complaints from neighbours or due to local traffic congestion impacting on the reliability of deliveries.

Next, appropriate resources need to be available. These 'policy resources' include not only money (financial) and personnel (human) input, but also law (legal), information (cognitive), organisation (interactive), consensus (confidence), time (temporal), infrastructure (property) and political support (majority) (Knoepfel et al 2007). Less directly, the culture of the organisation, current financial position, overarching goals and the structure of an organisation will have an impact in some way (see Chapter 5 for more on this).

Finally, taking a further step back, the political, economic, social, technological, legal and environmental conditions of the surrounding area will play a significant role. So, for example, a site located in a deeply rural area or on an edge-of-town site will have very different access and mobility needs to one occupying an address in the centre of a major metropolitan city, whilst the financial and political climate may influence an organisation either to engage with or ignore the Travel Plan agenda.

4.6 Travel Plans as a Package of Transport Instruments.

The final point concerns the package of instruments offered by the Travel Plan – that is the sharp end.

As with transport authorities of whatever level formulating and implementing transport plans, the holy grail for Travel Planners is to devise a strategy that is effective, popular with the decision makers and the public and yet inexpensive in terms of resources – hence the need for a 'package'.

In practice, a host of good practice guides propose a full range of available instruments for the Travel Planner to 'plug and play' (see Table 2.11). However, once the 'contextualisation' warnings have been observed, in reality the selectable instruments in a particular case are probably rather less.

Table 2.11 Travel Plan measures contextualised

Type of Measure	Examples	Contextual constraints			
		Effectiveness	Cost	Acceptability	Practicality
Supply-led measures	Bus stop improvements and/or enhanced frequencies	Low	High	High	May need bus operator/LA involvement
	Cycle parking, lanes and so on	Low	Medium	High	Only useful if site near cycle network
	New car parking spaces and/or access roads	High	Very high	High	Space required
Fiscal incentives	Incentives to use bus, rail, cycle, car share	Medium	Very high	Very high	Only useful if good PT access
Fiscal disincentives	Introduce parking fees	Very high	Low	Very low	Politically difficult
	Alter mileage rates	High	Low	Low	Politically difficult
Management	Ration parking spaces	Very high	Low	Very low	Politically difficult
	Alter business travel rules	High	Low	Low	Politically difficult
	Teleworking/ videoconferencing	Medium	Medium	Variable	Dependent on industry type
	Homeworking, compressed working weeks	Medium	Medium	Variable	Dependent on industry type
Information and marketing	Improve bus, rail, cycle info.	Very low	Low	High	Needs bus operator/LA involvement

Simplistically, (as alluded to in Chapter 1 for transport measures generally):

- Low cost measures such as improving information or installing secure cycle parking are generally not very effective at addressing transport problems but are acceptable.
- Measures effective at meeting transport objectives are usually unpopular to introduce and operate, but are sometimes cost neutral or even raise money.
- Measures that are popular such as public transport fare discounts or improved bus service levels are expensive and may only be of limited effectiveness in meeting transport objectives.

In addition, expensive measures can generally only be introduced by larger organisations; measures such as videoconferencing and homeworking will only be suitable for certain sectors; public transport and cycle improvements are only viable where existing networks can be added to; and often measures can only be introduced with the cooperation of other stakeholders such as local authorities and/or transport operators. As can therefore be appreciated, things are not always as straightforward as they might be.

5 How Well do they Work?

Travel Plan performance can most usefully be reported at the site level and at the wider regional and national network levels. The following sections first consider Travel Plan performance at the organisational level, and then at their impact on society more generally. In particular, the performance of Travel Plans against transport goals and then against wider organisational and/or societal aspirations are examined. Lessons learnt from the literature are then reported.

5.1 Site Level Impacts

At the site level, Travel Plans can deliver significant transport (and wider) benefits (see Table 2.13). In the UK, Cairns et al (2005) reports results of 24 case studies of Workplace Travel Plans carried out in 2004 that indicated reduced car use of between 10 per cent and 25 per cent, whilst the range for School Travel Plans is 8 per cent to 15 per cent and 7 per cent to 15 per cent in urban areas for Personalised Travel Planning. Next Hillsman et al (2001) evaluates the Commute Trip Reduction (CTR) programme in Washington State and finds the programme reduced single occupancy vehicle trips at participating sites from 75 per cent in 1993 to 68 per cent in 1999 – a drop of 6 per cent. Meanwhile the average reduction in trips from the Smarter Travel Workplaces Programme in Ireland is 18 per cent (NTA 2010).

In cost terms, in the UK DfT and NBTN (2008) cites a figure of £GBP50 (€56) per employee, and Cairns et al (2008) reports a median average of £GBP47 (€53) per employee for implementing a Travel Plan. Rather lower, Irish experience suggests costs of €7 per employee engaged and a further €2 for ongoing maintenance (NTA, 2010). Using a different measure, Rye (2002) calculates a cost per trip removed from the road network of €1.50.

Table 2.12 Effectiveness of Travel Plans at the site level

Area/Sample	Method	Transport effectiveness	Financial effectiveness	Source
24 Workplace Travel Plans, UK	Case studies	10 per cent – 25 per cent reduction in SOV		Cairns et al (2005)
School Travel Plans, UK		8 per cent – 15 per cent reduction in SOV		Cairns et al (2005)
Personalised Journey Planning (urban areas, UK)		7 per cent – 15 per cent reduction in SOV		Cairns et al (2005)
CTR programme in Washington State		SOVs from 75 per cent to 68 per cent (an average 6 per cent fall at participating sites)		Hillsman et al (2001)
Smarter Travel Workplaces Programme, Ireland		Reduction in SOV trips of 18 per cent	€7 per employee engaged plus €2 for ongoing costs	NTA (2010)
UK	Case studies		£GBP47 per employee (€53)	Cairns et al (2008)
UK	Case studies		£GBP50 per employee (€56)	DfT and NBTN (2008)
UK	Case studies		€1.50 per trip removed from the road network	Rye (2002)

Overall, then, the evidence suggests that car trips can be cut by up to 30 per cent in exceptional circumstances, but that a reduction of between 5 per cent and 15 per cent is more usual at the site/organisational level.

5.2 Regional and National Impacts

At the regional level, the evidence has not been quite so dramatic. Dill (1998) finds that regional reductions in car trips of 4 per cent to 5 per cent are commonly cited but are still not seen as sufficient by a business sector complaining about high costs and bureaucratic hassle. Pansing et al (1998: 101) compares 58 measures and finds 'organisational TDM' to be "relatively effective and cost effective" at reducing vehicle trips compared to other measures (though based on a limited sample).

At the regional level in the USA, Loudon et al (2003) reports that combining a range of TDM strategies for employment and residential case study areas along Corridor SR520 in Washington State resulted in an overall reduction in peak period commute trips for the afternoon of 6 per cent, which translated into 0.2 per cent of total eastbound traffic and 0.3 per cent of total westbound traffic on the corridor as a whole. Bianco (2000) reports that a comprehensive programme of TDM measures in a commercial district of Portland, Oregon reduced single occupancy vehicle trips by 7 per cent to 56 per cent. Of the sample surveyed who changed their travel behaviour, a quarter changed due mainly to reasons unconnected with the measures, 22 per cent due to new parking metres and 19 per cent due to the public transport incentive scheme. Similarly, Hillsman et al (2001) estimates that traffic fell by less than 1 per cent across the Puget Sound region during the morning commute for an annual investment of $USD3.1 million (€2.1 million) from Washington State plus local matching funds from local councils and employers as a result of the State-wide CTR programme. And, in a modelling study of Syracuse, New York Taylor et al (1997) estimates that comprehensive employer TDM programmes would deliver a reduction of 14 per cent to 18 per cent of vehicle trips (assuming 75 per cent of employees participate in the sub-area), whilst the figures would be 13 per cent to 16 per cent if 55 per cent of employees participate. The corresponding ranges for reductions in Vehicle Kilometres Travelled (traffic) would be 13 per cent to 17 per cent and 12 per cent to 15 per cent respectively.

Looking at the national level, Hoogma (2006) reports that in the Netherlands of roughly 16,000 companies with more than 50 employees, 30 per cent have been directly informed of Travel Plans, while 6 per cent have implemented them. On average, the reduction in car use realised has been 8 per cent. Despite this, though, in the view of ECORYS-AVM (2005: 5) there is frustration that company mobility management results after more than ten years "often did not have the desired result" at the [national (Dutch)] network level. In Ireland, NTA (2010) notes that currently 56,000 staff in Dublin have been engaged in Travel Planning and that the target is to reach 250,000 employees (14 per cent of employees nationally) by 2014. Meanwhile, in the UK, it would seem that Travel Plan measures remain a marginal tool among businesses. BCC (2007) reports that only 6 per cent of businesses have a formal written Workplace Travel Plan, while 4 per cent public transport incentive schemes, 6 per cent have reduced parking spaces, 13 per cent have introduced car sharing schemes and 13 per cent walking/cycling schemes. More encouragingly, 44 per cent now allow staff to work from home, 40 per cent use IT to reduce business travel and 34 per cent have adopted staggered work hours or flexi-time.

Earlier work in the UK by Bradshaw and Camara (1999) reports that only about 3 per cent of local authorities had implemented a Travel Plan for their own employees on a permanent basis, although a further 4 per cent had implemented one on a pilot or trial basis (for example, covering only some departments), while a further 40 per cent had started the process. At the same time about 6 per cent of authorities had encouraged Travel Plans at other organisations in their area on a permanent basis, a further 6 per cent had encouraged pilots at other organisations and 23 per cent had started the process by contacting employers. Rye (2002) estimates that the average effectiveness of Travel Plans per organisation in reducing single occupancy commuting trips to be 6 per cent. He then scales up this figure to the Great Britain network using national statistics and calculates that Travel Plans reduce the amount of car use by 3.7 million kilometres per day, or the equivalent of 0.6 per cent of overall car traffic.

In terms of cost effectiveness, Van Hattum (2004) draws on US data to show the cost of accommodating one additional single occupancy vehicle trip by expanding roads to be $USD1,258 p.a. (€870) compared with $USD530 p.a. (€365) using TDM strategies.

Table 2.13 summarises these results.

Table 2.13 Regional and national level Travel Plan effectiveness

Area/sample	Method	Transport effectiveness	Financial effectiveness	Source
Worldwide	Literature review	2 per cent to 5 per cent shift from car	Businesses see 'high costs and bureaucratic hassle'	Dill (1998)
USA	Compares 'limited sample' of organisational TDM measures with other Transport Control Measures	'Relatively effective'	'Cost effective'	Pansing et al (1998)
Corridor SR520 Washington State		6 per cent fall in pm peak period trips, or 0.2 per cent reduction in all eastbound and 0.3 per cent fall in westbound traffic		Loudon et al (2003)
Comprehensive TDM package in commercial area of Portland, Oregon	Survey	Reduced single occupancy trips by 7 per cent (quarter of these changed for reason unconnected to TDM measures)		Bianco (2000)

CTR program in Washington State		Traffic fell by less than 1 per cent across Puget Sound region in am peak.	Hillsman et al (2001)
Netherlands	Survey	Car use reduced by 8 per cent	Hoogma (2006)
UK	Informed estimate of actual impact of Workplace Travel Plans	Reduction of 0.6 per cent in VKT	Rye (2002)

Overall, the figures suggest that regional and national level Travel Plan impacts are in the order of 0.2 per cent to around 8 per cent, depending on the degree of 'penetration' (level of take up) achieved. These results and their implications will be explored further in Chapters 5 and 6.

In explaining this reduction in effectiveness from the site level to wider areas, TCRP (2010) suggests that this is due to a dissipation of effects in a number of ways, namely:

- Effectiveness is levelled out amongst differing organisations, depending on their size and type of measures introduced.
- Results are mixed in with non-participating organisations and with non-targeted trips (leisure, education and shopping trips).
- Impacts are further diffused as locally generated trips are mixed on specific facilities with intra-regional and intercity travel (as on major highways for example).

6 Lessons Learnt

In learning lessons, once more it is useful to look first at the organisational level, and then at the wider community level.

6.1 Lessons for Organisations

At the organisational level, Hoogma (2006) notes that the arguments of better accessibility, more effective/cheaper use of parking spaces and environmental benefits have been largely successful, but that there are also problems. Specifically, he suggests that the Travel Plan concept is not clear; is seen as being time consuming; does not have a good image; is difficult to incorporate into collective labour agreements; and requires all organisations to participate (which does not often happen) if the 'free rider' issue is not to arise. Attenbrand et al (2005) adds that MM tools do not clearly fit into any traditional organisational segment; and says the direct benefits of MM initiatives are difficult to demonstrate and calculate (which is probably the reason why employers do not usually offer any transport services other than car parking or company cars which are supported and justified by building regulations and tax codes). Next, Bradshaw and Camara (1999) comments that more financial resources and more information are needed if such measures are to be adopted and be successful. ECORYS-AVM (2005) discusses the need for targets to be focused on the needs of the organisation and for the link between the problem and the approach to be made explicit. Finally, Rye (1997) draws on the development of five public sector Workplace Travel Plans – one in southern California, one in Hertfordshire, UK and three in Nottinghamshire, UK to conclude that the principal 'ingredients' of implementing a Travel Plan should be:

- A corporate Travel Plan working group.
- Alliances with other 'relevant' groups within the organisation.
- A senior member of staff to support the Travel Plan.
- Identified resources.
- An employee transportation coordinator.
- Modest, achievable targets set by departmental management.

From the evidence above, it is clear that Travel Plans are not seen as being integral to the workings of an organisation nor as something positive. Rather, they are perceived to be ill-defined and burdensome, and something that is applied by an external actor without regard as to how it actually might properly fit within the structure of the organisation (see Chapter 6 for more on this).

6.2 Lessons for Policy Makers

Linking these issues to policy, Ungemah and Dusza (2009) notes that the effectiveness of a TDM strategy varies depending on: the type and degree of strategies (measures), the affected area and the extent of concurrent supporting strategies. More specifically, Zuehlke and Guensler (2007) notes that practitioners and policy makers have struggled to influence employers to adopt Travel Plans, and that experience suggests that planning processes need to be far more inclusive when developing strategies and implementation goals. Meanwhile Zupan (1992) finds the following:

- Locally targeted strategies can relieve congestion in spot locations, but their impact on wider congestion is difficult to measure and thought to be small.
- Area-wide TDM strategies, rather than those covering specific locations or with narrowly drawn geography, have the most potential for congestion relief.
- Establishment of performance objectives, rather than a prescription of specific actions, inspires greater innovation and success.

- Voluntary actions are much less likely to lead to success than mandated actions.
- Economic self-interest inspires successful actions.
- Realistic alternatives to single occupancy vehicles must be present for changes to occur.
- No legal area-wide mechanisms have yet been devised in the absence of success through voluntary actions.

Finally, in a SWOT (strengths, weaknesses, threats, opportunities) analysis of Travel Plans from a Dutch perspective (see Table 2.14), Boot et al (2007) reports that Travel Plans are cheap and effective, but that this is often not the perception of policy makers. Actions are therefore required to alter this perception (see Chapters 6 and 7 for more on this).

Table 2.14 Travel Plan SWOT analysis

Strengths	**Weaknesses**
Alternative means of transport are required due to environmental problems, increasing fuel prices, large road maintenance projects and pricing policy; Obligation of mobility management (environmental law); Political attention for MM; Many initiatives, certainly in the Randstadt area.	Use of regional public transport is decreasing; Good mobility managers are scarce; Companies with under 200 employees do not pay enough attention to mobility management; Business trips with your own car are an interesting means of earning money.
Opportunities	**Threats**
Much can be reached with a small budget; MM is cheaper than new infrastructure; MM connects transport policies with environmental and health policies; Reduction of space required for parking; Reduction of mobility expenses.	Perception that MM is not effective (although this is changing); It is difficult to get budget for small projects; MM sometimes lacks focus; Lack of priority at small municipalities.

Source: Boot el al (2007)

7 A Suggested Analysis Framework

From this chapter it can be seen that the Travel Plan in its current guise consists of six key elements, namely:

- Travel Plans are not really an instrument themselves, but a delivery mechanism or strategy for other mostly transport-focused measures.
- Travel Plans are delivered by an additional 'agent' that is not a part of the 'traditional' transport policy institutional structure.

- Travel Plans are initiated in two ways: either by the organisation or by government.
- Travel Plans seek to deliver transport and related benefits to the community in addition to some more direct organisational benefits to the participating 'delivery agents'.
- Travel Plans are to some extent 'site-specific' and tailored to the specific contextual circumstances.
- Travel Plans to some degree deliver a package or a strategy of a 'wide variety' of transport instruments.

Moreover, these can be combined to provide the definition that Travel Plans are a mechanism for delivering a package of transport measures targeted at a specific site by an agent with a strong relationship with the local transport users to deliver transport, and wider goals to the organisation and society as a whole.

Overall it would seem that while relatively successful as a transport strategy at a neighbourhood level, so far at least this success has not been replicated at the national or even regional scale. To explore why this has been the case so far, and to examine some possible future scenarios, the following chapters will each consider one so-called 'axis of Travel Plan development'. This model was first suggested in Enoch and Potter (2007) but is here substantially revised and expanded upon in the following chapters (see Figure 2.3).

From this, the development of Travel Plan segments from workplaces, to schools, events, attractions, retail outlets, transport interchanges, residential developments, individuals (through personal journey planning) and freight operations will be explored in Chapter 3. Next, Chapter 4 investigates issues of scale, and presents a range of approaches to organising groups or networks of Travel Plan organisations, from informal to more formalised, as a means of delivering Travel Plans more effectively. Chapter 5 (structure) then explains how Travel Plans are evolving internally within organisations in terms of motivations, degree of involvement, timing of intervention, the information process involved and the delivery agents used. Chapter 6 then focuses on the issue of 'support'. Particularly, it first examines how Travel Plans began in current organisations on an ad hoc basis but are now being increasingly 'pushed' by public authorities through awareness raising activities, capacity building exercises, regulations, subsidies and tax changes, before looking also at institutional measures, indirect support approaches and issues of leadership.

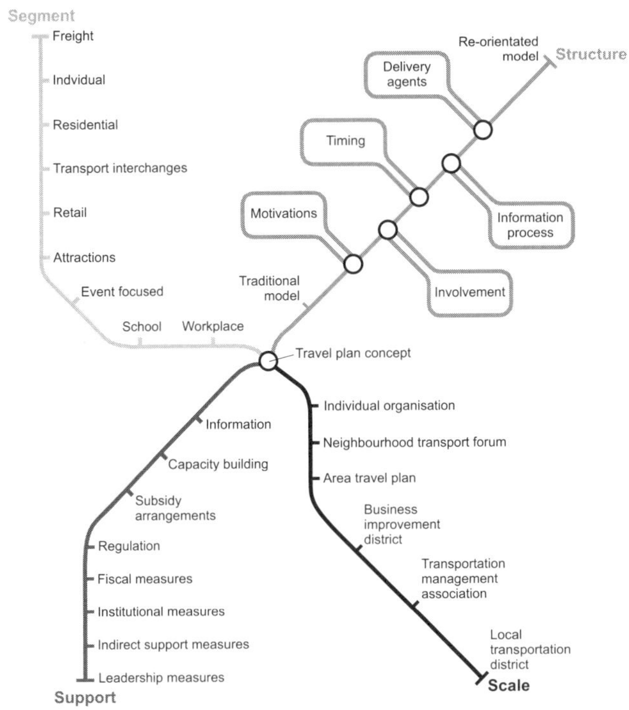

Figure 2.3 Travel Plan axes – segment, scale, structure and support

Telecommunications company Orange provides works buses for its employees in Bristol

In the mid-1990s, Addenbrookes Hospital in Cambridge was one of the first UK organisations to adopt a Travel Plan

References

Attenbrand, A. S., Jorde, B., Kasin, O., Krag, T, Silverberg, B., Skur, J. and Stenvall, M. 2005. *Mobility Management in the Nordic Countries*, TemaNord, 539, (Copenhagen: Nordic Council of Ministers).

Berman, W., Differt, D., Aufschneider, K., DeCoula-Souza, K., Flemer, P., Hoang, L., Hull, R., Schreffler, E. and Zammit, G. 2006. *Managing Travel Demand: Applying European Perspectives to US Practice*, FHWA-PL-06-015, Federal Highways Administration, US Department of Transportation, May. (Washington DC: FHWA).

Bianco, M. J. 2000. Effective transportation demand management: combining parking pricing, transit incentives, and transportation management in a commercial district of Portland, Oregon, *Transportation Research Record,* 1711, 46–54.

Boot, R., Metz, F. and Pauwels, H. 2007. State of the art paper on mobility management in the Netherlands. *European Platform on Mobility Management.* [Online]. Available at http://www.epomm.org [accessed 14 October 2009].

Bradshaw, R. and Camara, P. 1999. The future of mobility management in the UK. Paper to the 27th Annual Meeting, *European Transport Conference*, Cambridge, 27–29 September.

British Chambers of Commerce. 2007. *Running on Empty: A Transport Survey*, (London: British Chambers of Commerce).

British Standards Institute, 2008. *National Specification for Workplace Travel Plans*, Publicly Available Specification 500, (London: British Standards Institute), November.

Cairns, S., Sloman, L., Newson, C., Anable, J., Kirkbride, A. and Goodwin, P. 2005. *Making Smarter Choices Work*, (London: Department for Transport).

Cairns, S., Sloman, L., Newson, C., Anable, J., Kirkbride, A. and Goodwin, P. 2008. Smarter choices: assessing the potential to achieve traffic reduction using soft measures, *Transport Reviews*, 28(5), 593–618.

Comsis Corporation. 2002. *Public Agency Guidance on Employer-Based TDM Programs*. Prepared for Transit Cooperative Research Program Project B-4, TCRP Web Document 22, Contractor's Final Reports, December. (Silver Spring: Comsis Corporation).

Darbera, R. 2002. Company management of staff's commuting choices: an economic view, in *Managing Commuters' Behaviour: A New Role for Companies*, edited by European Council of Ministers of Transport, Round Table 121 (Paris: ECMT/OECD), 237–251.

DeHart Davis, L. and Guensler, R. 2005. Employers as mediating institutions for public policy: the case of commute options programs, *The Policy Studies Journal*, 33(4), 675–697.

Department of the Environment, Transport and the Regions (Great Britain). 1998. *A New Deal for Transport: Better for Everyone*, (London: The Stationery Office).

Department for Transport (Great Britain). 2007. *Towards a Sustainable Transport System: Supporting Economic Growth in a Low Carbon World*, (London: Department for Transport).

Department for Transport (Great Britain) and National Business Travel Network. 2008. *The Essential Guide to Travel Planning*, (London: Department for Transport).

Dill, J. 1998. Mandatory employer-based trip reduction, *Transportation Research Record*, 1618, 103–110.

ECORYS-AVM. 2005. *Where there is a Will, there is an Effect: Experiences and Case Studies with Mobility Management in the Netherlands*, (Rotterdam: KpVV).

Energy Efficiency Best Practice Programme. 2001. *A Travel Plan Resource Pack for Employers*, (London: The Stationery Office).

Enoch, M. P. and Potter, S. 2007. Travel planning, in *Managing Transport Energy: Power for a Sustainable Future*, edited by J. P. Warren, (Oxford: Oxford University Press), 123–167.

Ferguson, E. 1997. Privatisation as choice probability, policy process and programme outcome: the case of transportation management associations, *Transportation Research Part A*, 31(5), 353–364.

Fishback, P. V. 1992. The economics of company housing: historical perspectives from the coal fields, *Journal of Law, Economics and Organisation*, 8(2), 346–365.

Flynn, C. P. and Glazer, L. J. 1989. Ten cities' strategies for transportation demand management, *Transportation Research Record*, 1212, 11–23.

Hendricks, S. 2008. Four challenges to incorporating transportation demand management into the land development process, *Transportation Research Record*, 2046, 30–36.

Herzog, E. and Grant, M. 2002. Commuter choice benefits calculator: web-based tool for estimating costs and benefits of commuter programmes, *Transportation Research Record*, 1781, 32–39.

Hillsman, E. L., Reeves, P. and Blain, L. 2001. Estimation of Washington State's trip-reduction programme on traffic volumes and delays: central Puget Sound region, *Transportation Research Record*, 1765, 16–19.

Hoogma, R. 2006. Company/commuter mobility management in the Netherlands. Paper to the 5th Annual Event for Energy Management Agents in Europe, *European Conference on Local Energy Action*, Brussels, Belgium, 7–8 February.

Knoepfel, P., Larrue, C., Varone, F. and Hill, M. 2007. *Public Policy Analysis*, (Bristol: Policy Press).

Loudon, W. R., Luther, D. K., Mabry, J. and Kavage, S. 2003. Travel demand management and land use effectiveness review for the Puget Sound region. Paper to the 82nd Annual Meeting, *Transportation Research Board*, Paper 03-3337, Session 339: Estimating the Impact of Innovative Transport Demand Management Strategies, Washington DC, 12–16 January.

Lund, B. 2006. *Understanding Housing Policy*, (Bristol: Policy Press).

Martz, J. 2006. *The American Experience*. Paper presented to the Association of Commuter Transport Conference, London, 21 February.

MAX Consortium. 2007. Definition of Mobility Management and categorisation of Mobility Management measures, MAX: successful travel awareness campaigns and Mobility Management strategies, Annex E to the Comprehensive Research Plan, Project 518368, European Commission Sixth Framework Programme, Brussels. [Online] Available at http://www.max-success.eu/ [accessed 24 June 2010].

Miller, J. H., Hough, J. and Hellegaard, L. 2005. Asset management and funding: transport and non-profit institutions, in *Handbook of Transport Strategy, Policy and Institutions,* edited by K. J. Button and D. Hensher, Handbooks in Transport: Volume 6, (Amsterdam: Elsevier), 391–402.

Ministry of Transport, Public Works and Water Management (Netherlands Government). 1992. *Transportation Demand Management: For Companies with Foresight*, V&W/DGV/022/92, (The Hague, Ministry of Transport, Public Works and Water Management).

MOST, 2008. Mobility Management strategies for the next decades (MOST), Summary of the European Commission 5th Framework Programme for the Transport Knowledge Research Centre. [Online] Available at http://www.transport-research.info [accessed 20 March 2011].

National Transport Authority (Republic of Ireland). 2010. Smarter Travel Workplaces, National Transport Authority, Dublin, Ireland. [Online] Available at http://www.smartertravelworkplaces.ie [accessed 18 March 2011].

Pansing, C., Schreffler, E. N. and Sillings, M. A. 1998. Comparative evaluation of the cost-effectiveness of 58 transportation control measures, *Transportation Research Record,* 1100, 97–104.

Rye, T. 1997. *The Implementation of Workplace Transport Demand Management in Large Organisations*, PhD Thesis, (Nottingham Trent University: Department of Civil and Structural Engineering).

Rye, T. 2002. Company management of staff's travel choices in the UK, in *Managing Commuters' Behaviour: A New Role for Companies*, edited by European Council of Ministers of Transport, Round Table 121 (Paris: ECMT/OECD), 175–200.

Schreffler, E. N. 1986. Transportation management organisations: an emerging public/private partnership, *Transportation Planning and Technology*, 10(4), 257–266.

Steiner, R. L. 1992. Lessons for transportation demand management from utility industry demand-side management, *Transportation Research Record*, 1346, 14–17.

Taylor, C. J., Nozick, L. K. and Meyburg, A. H. 1997. Selection and evaluation of travel demand management measures, *Transportation Research Record*, 1598, 49–60.

Thaler, R., Herry, M. and Knoflacher, H. 2002. Successful approaches to environment and economy – company management of staff's travel choices, in *Managing Commuters' Behaviour: A New Role for Companies*, edited by European Council of Ministers of Transport, Round Table 121 (Paris: ECMT/OECD), 5–40.

Traject, Studio Ripamonti, TC&O Management voor Mobiliteit, VIA Beratende Ingenieure, Leeds City Council and Austrian Mobility Research AMOR. 2000. Toolbox for mobility management in companies, Contract XVII/4.1031/Z/98-018, Directorate General XVII (Energy), SAVE II Programme, European Commission, Brussels. [Online] Available at http://www.mobilitymanagement.be/index.htm [accessed 24 June 2010].

Transit Cooperative Research Program. 2010. Employer and institutional TDM strategies, Chapter 19 of *Traveller Response to Transportation System Changes*, TCRP Report 95, (Washington DC: Transportation Research Board).

Ungemah, D. and Dusza, C. 2009. A transportation demand management (TDM) benchmark: results from the 2008 TDM program survey. Paper presented to the 88th Annual Meeting, *Transportation Research Board*, Paper 09-2174, Session 441: Assessing Traveller Motivation and Need for Transportation Demand Management Applications, Washington DC, 11–15 January.

Van Hattum, D. 2004. Expanding commuter options in the Twin Cities, *Minneapolis-St Paul: Minnesota Sustainable Communities Network*. [Online] Available at http://www.nextstep.state.mn.us/download/expandingcommuteroptions.pdf [accessed 17 October 2009].

Wilhelm, A. and Posch, K. H. 2003. Mobility management strategies for the next decades, *Transportation Research Record,* 1839, 173–181.

Winters, P. 2010. History of carpooling and vanpooling as we know it. Paper presented to the 89th Annual Meeting, *Transportation Research Board*, Session 151, Reinventing carpooling to meet transportation's greatest challenges, Washington DC, 10–14 January.

Zuehlke, K. and Guensler, R. 2007. Employer perceptions and implementation of commute alternatives strategies, *Journal of Public Transportation*, 10(4), 171–194.

Zupan, J. M. 1992. Mandatory employer-based trip reduction, *Transportation Research Record*, 1346, 1–9.

Chapter 3
Segment: Examining the Influence of Activities on Trips and Travel Plans

1 Travel Plan Segments

The first Travel Plan development axis to be examined is that of 'segment'. Specifically, workplaces will be considered first, followed by schools, planned events, attractions, retail outlets, transport interchanges and residential developments. Finally, for completeness, personal journey plans and Travel Plans for freight trips will be covered, though it could be argued that neither are Travel Plans in the strict sense of the term. Within these categories, it is instructive to consider four types of characteristic – those of the organisation, the location, the user and the trip or journey (which are effectively a derivative of the others). The implications for the design, operation and performance of the Travel Plan in each circumstance will then be discussed.

1.1 Organisation Characteristics

Crucial organisational factors that impinge on transport outcomes include the size (in terms of numbers of employees, turnover, available resources and so on); their motivations, objectives and their users, as well as the wider context in which they operate and the resources that they have available. In general, it can be assumed that larger organisations (see Chapter 4) with strong motivations for needing to change people's travel behaviour, sufficient resources (in terms of finance, suitable skills, personnel, time, management commitment), and a high degree of control over those accessing their site will be in the best position to introduce and operate an effective Travel Plan.

1.2 Location Characteristics

Clearly the physical location of an organisation directly influences the travel behaviour of those accessing it. Typically, 'activity intensive' locations (usually city centres accommodating mainly shops and commercial development) generate a high demand for transport and so consequently are well served by public transport. This high level of accessibility is reflected in high land values, which in turn ensures that the intensity of activity (and hence the high levels of demand for transport) are maintained. By contrast, edge of town and rural areas, with low levels of activity, typically do not generate sufficient levels of demand to justify more than a limited public transport service which in turn results in lower land values and land uses that are commensurate with that status (agriculture, industrial parks, out of town retail developments). Simplistically, then, in city centres public transport access is good, car access is difficult due to congestion and car parking is expensive, meaning that public transport use is high and car use is low. In rural areas, public transport access is poor, car access is easy and car parking is free, meaning that public transport is low and car use is high. Denser development and other environmental factors generally means that walking and cycling are more likely to be a realistic option for people in urban than in rural areas.

1.3 User Characteristics

The type and location of the organisation also affects the characteristics of the users. Possible types of user include: employee, visitor, supplier, client/customer, pupil/student, patient and spectator/worshipper. Of these, the groups where most information useful to Travel Planning is generally known are employees, pupils/students, suppliers and sometimes patients, whilst a lot or almost nothing may be known about the remaining groups depending upon circumstances. It could also be assumed that the level of such information known about those groups relates to the level of influence exerted by the organisation on the individual, and hence onto the potential effectiveness of the Travel Plan measures in changing their travel behaviour.

Looking in more detail at employees, it is the working conditions (particularly arrival and departure times) that most influence the Travel Plan-relevant attributes of an organisation. Similarly, pupils and students are required to attend classes at fixed times and locations. Meanwhile, the transport patterns of suppliers are also trips over which an organisation can potentially exert a high degree of influence on the types of transport access allowed to sites through the use of procurement contracts.

In many situations the difference between customers and visitors is somewhat blurred. For example, football fans may well be both at the same time. However, one difference might be that visitors are more often tied to a particular destination (for example, a hospital to see a sick relative) or type of destination (perhaps a historic town, museum or national park), while customers first seek the 'product' to buy and then choose from a range of locations of where to buy it.

In principle, then, this 'destination limitation' would seemingly make the visitor an easier target for Travel Planners than the more footloose shopper. But in practice this is only part of the story because there are also (at least) two other dimensions, namely 'intensity' (length and frequency of the relationship) and 'level of information' offered (perhaps through a loyalty card or home delivery services). It is here that certain types of shopper potentially become the more attractive Travel Plan market (see Figure 3.1).

This is because while the relationship intensity will vary enormously between individuals and over time, it is retailers more than visitor 'attractions' that tend to collect the higher levels of information about their clients. This is done through loyalty cards and/or home delivery options (particularly supermarkets) from which highly detailed pictures of the preferences of individuals can be formed, based on customer buying patterns, especially where trips are high frequency and made over a period of time. This knowledge can include exactly where they live, their likely trip frequencies, and whether, and even how far (based on petrol sales) they drive, meaning that, in theory, such organisations could draw on such information to develop individually tailored incentives to affect how these people accessed the site.

However, organisations looking to sell their products face high levels of competition meaning that they are often unwilling to consider any measures such as car parking restrictions or charges at the retailing point that may place them at a competitive disadvantage.

Finally, hospital patients can either be well suited to Travel Plan measures or not well suited depending on whether they arrive in an emergency or not (see later).

In addition, individual characteristics such as age, gender, income, educational attainment, socio-economic status, home location, cultural influences, status in the family and whether or not someone owns and/or has access to a car will influence people's travel behaviour and thus their propensity to respond to Travel Plan measures. Thus, Anable (2005) draws on behavioural theory to note the importance of transport planners segmenting the travel market so as to focus resources on convincing the most susceptible groups of people to reduce their use of the car, while Balcombe et al (2004) summarises a wealth of previous (mostly economics-based) studies to outline the individual factors affecting public transport use.

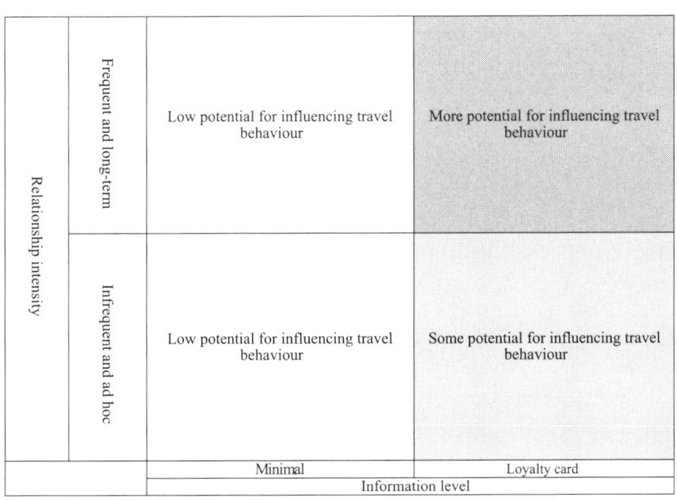

Figure 3.1 Identifying the market for influencing travel behaviour

1.4 Trip Characteristics

Given the truism that transport is often considered to be a derived demand, then the characteristics of the trip are dependent on the factors already discussed previously. Thus, journey purpose stems from the type of organisation and the user. Other trip characteristics relevant in the Travel Plan context include the frequency that the trip is made (which relates to the familiarity of the trip to the user), time of travel (whether peak or off-peak), individual or group, amount of luggage, importance of journey reliability and the cost of the trip. In addition, of crucial importance in Travel Planning terms is the degree of control exercised by the organisation over the individual for particular trip purposes. So, for example, evidence from Haifa, Israel suggests that the dominant influences on mode choice for commuting trips are car ownership and car availability, but that employment site characteristics (for example, distance of site from central business district, employment density, percentage of industrial land and number of peak bus services to the site) are also significant (Shiftan and Barlach 2002).

Looking at the trip characteristics in more detail, DfT (2010) provides a useful list of journey purposes, which include commuting, business, education, shopping, personal business and leisure, and these will be used as a basis for the next section. Before that, however, Table 3.1 draws in data from the same reference for the UK case to provide an indication of the relative importance from a Travel Planning perspective of each of the journey purposes.

Table 3.1 UK journey purpose share by number and length of trip

	Proportion of trips by journey purpose		**Proportion of trip length by journey purpose**	
Purpose	Per cent of trips by journey purpose	Per cent for each purpose by car	Per cent of trips by journey purpose	Per cent for each purpose by car
Commuting	**15**	67	**19**	73
Business	**3**	78	**8**	78
Education/escort	**11**	42	**4**	53
Shopping	**20**	63	**12**	82
Other escort	**9**	84	**7**	94
Personal business	**11**	65	**7**	82
Leisure	**27**	69	**42**	80
Other	**4**	1	**1**	13
All purposes	**100**	63	**100**	78

As can be seen from Table 3.1, commute trips make up roughly a fifth of car traffic levels. Moreover, they typically occur on a regular (usually daily) basis during peak periods, are made by individuals and do not usually involve much luggage being carried, often making them an ideal target for public transport and/or car sharing initiatives. Reliability is important, and the cost (which can be significant) is borne either by the commuter (for example in the UK, Ireland and the USA) or at least partially by the employer too (as in Germany, Switzerland and the Netherlands) (Potter et al 2006). In total, this means that the employer can have some degree of control over the commuting trip, and consequently, commute trips have tended to be the major target of Travel Plan measures over the years.

Rather less often addressed by Travel Plans is business travel, which Alamgir (2010) characterises as being:

- trips by passenger and not freight;
- travel by employees and not others;
- travel that is related to the purpose of the organisation and not personal travel; and
- travel not including trips to and from home.

This lack of attention may seem strange given that responsibility for (and the associated costs of) 'travel in the course of work' or 'corporate travel' is entirely within the purview of an organisation – that is the degree of control by the organisation over the individual is high – whereas, as noted previously, in many countries commuting is often the responsibility of the employee. Perhaps more pertinently, at the organisational level Morrison et al (1994) states that several US companies spent more than £GBP100 million (€112 million) a year on business travel (and that in the early 1990s). More recently Global Expense (2008) estimates that in 2007 roughly 3.75 million people (of 30 million in the UK workforce) made expense claims of £GBP1,555 (€1,750) on average, of which some 49 per cent were for business travel – which equates to £GBP5.8 billion (€ 6.5 billion) altogether.

Overall, such a lack of attention by transport policy officers would seem to be due to the fact that business travel typically accounts for a far smaller proportion of travel than commuter journeys in terms of trips (3 per cent versus 15 per cent in the UK). However, once average journey distances are incorporated the respective proportions are 8 per cent and 19 per cent respectively, while car use levels are also proportionally higher for business travel than commuting (see Table 3.1 and DfT 2010) meaning that such a rationale is not as robust as it might first appear. Interestingly, business travel has also been largely ignored by the academic community (Beaverstock et al 2010).

Here the characteristics are that the trip is infrequent and so is unfamiliar, and often does not take place at peak times. Instead, travel options need to be flexible and conveniently timed. Like commuting, business travel is usually undertaken by individuals not groups and luggage is not an issue, though reliability is usually more important than for commuting, while the cost constraint for business travellers is reduced (because the costs incurred are paid for by the employer). One other point is that business travellers tend to appreciate being able to work over the course of their journey – so comfortable seats and tables and access to communication technologies can be seen as important elements. Finally, for some trips (such as internal meetings) there is scope to substitute trips with virtual meetings (such as by using videoconferencing and teleconferencing) although for other trip types (like meeting clients) then face-to-face meetings remain highly valued (Swarbrooke and Horner 2001). Overall, then, in theory, business travel for employees looks to have high potential as a trip type to target for Travel Planners though organisational barriers may limit this market in the short-term.

Education trips occur at peak times between fixed origin and destinations. They are also characterised by the relative importance of non-consumers of the trip (namely parents). This is both as escorts of trip making students and as primary decision makers in terms of trip origin and destinations, mode choice and route assignment – though parents are to some degree influenced by their children as to how they behave, meaning that any school Travel Plan strategy would need to target both groups, (parents and students) to be effective (Van-Ristell 2011).

Trips to health facilities for medical treatment can be classified as being emergency or non-emergency. By their nature, emergency patient trips are clearly not susceptible to change, due to decisions being based on how best to deliver the required level of medical assistance within a set time. More promising is the non-emergency patient group. Here, medical centres will have information about the trip makers and at a minimum could provide them with tailored travel details, while the relative steadiness of demand over the course of the day could theoretically allow for appointments to be arranged to fit transport alternatives to the car or vice versa.

The term 'personal business trips' provides a useful catch all category to cover those trips such as journeys to the bank or the dentist, not referred to as health, shopping, leisure or work-related. While clearly varied in nature (potentially making them difficult for Travel Planners to target), Cairns (1997: 2) sees such trips as being "of a shopping nature", in that they involve "visiting a specific location to purchase or use a service or a good". Her paper adds that despite their significance in terms of trips made (11 per cent in the UK according to DfT 2010), such trips are rarely studied.

On shopping trips, Cairns (1997) sets out two major types, namely *convenience* shopping, which tends to be regular and mundane sorties for food; and *comparison* shopping, namely less frequent forays for other items, where looking around is important and there is an associated leisure component. In the UK, ONS and DfT (2007) reports that on average people made 206 trips to or from the shops in 2005, travelling 1406 kilometres. Of these, approximately half of all shopping trips (105) are for food (convenience) and the remainder are for non-food items (comparison). Food shopping trips tend to be shorter (average five kilometres) than non-food shopping (average eight-and-a-half kilometres), and over a fifth of shopping trips are undertaken on a Saturday – the most common day for shopping.

As noted in the previous section, retailers can either hold quite comprehensive dossiers on customers, or else know nothing at all. Typically trip behaviour is obviously influenced by the transport alternatives offered to sites (ranging from excellent in city centres to poor at certain edge and out-of-town sites), but also by the 'bulkiness' of the items bought. In addition, some shopping trips are perceived to require the shopper to be physically there in order to ensure that the right price is paid for the right quality of product. Thus, as with business trips there are certain types of shopping trips which are increasingly subject to being substituted by virtual trips (though the product then still needs to be physically delivered), and some which are far less so. Cairns (2005) finds that with realistic levels of take-up then the direct substitution could cut car traffic by 70 per cent or more, although this effect will be reduced once a variety of complex behavioural responses have been worked through. She adds that benefits could be further enhanced by the "use of appropriate cost structures, new types of delivery location, less polluting vehicles, greater cooperation or out-sourcing by retailers, and measures to encourage greater consumption of local produce" (Cairns 2005: 51).

"Leisure travel differs from utility travel in the degree of discretion exercised by the traveller in whether or not to travel, where, when and how to go" (Guiver et al 2008: 144), and as seen in Table 3.1 leisure trips are already the most important journey purpose in terms of impact, and indeed are the fastest growing. Yet, despite this, UK Government interest in addressing the tourist transport problem is not seen as being as high as for commuting or the 'school run' due to it being more variable and ad hoc, and hence more 'intractable' (Dickinson and Robbins 2008).

Finally, freight journeys to and from organisations (supplier, service and delivery trips) are a trip type which is sometimes considered by Travel Plans, though not always in the same process as for passenger trips. This is partly due to a fairly strong divide between passenger and logistics professions and probably because freight journeys are often already an integral part of organisational processes in a way that person trips are not. In the UK, Cherrett and Triantafyllou (2009) notes that *Delivery and service plans* and *Construction logistics plans* are broadly analogous to Travel Plans, with the emphasis being on increasing the efficiency of freight transport movements in each case, often by facilitating organisations working together where possible.

2 Workplace Travel Plans

Workplace or Employer Travel Plans are the most widespread worldwide, and indeed were the first to emerge (see Chapter 2). In categorising these further, Table 3.2 shows a grid with two axes – which denotes whether the organisation is from the private or public sector and the form of motivation (reason for adoption of Travel Plans).

Table 3.2 Travel Plan motivations for the public and private sectors

	Internal motivation (or voluntary)	External motivation (or imposed)
Private Sector	Type I	Type IV
Public Sector	Type II	Type III

Initially, Travel Plans (and their predecessors) emerged as a response to internal issues (albeit sometimes as a response to external events such as the fuel crises of the 1970s), predominantly of how to ensure that staff could get to and from the workplace. In the USA during this first 'Type I' phase it was generally private sector organisations that responded in this way, whereas later in European countries there has more often been a mix of private and public organisations (Types I and II in Table 3.2). Subsequently, the motivations to Travel Plans switched such that motivations external to organisations, generally in the form of dictats from national level bodies for public sector organisations like hospitals and local authority offices (Type III) and/or most commonly planning or employment regulations for private sector organisations (Type IV) began to predominate. Interestingly, however, there does appear to be a trend emerging in the UK and Ireland whereby private sector organisations are once more considering Travel Plans as a means of meeting internal 'Type I' objectives thus completing the circle (NTA 2010). In particular, the global recession has forced organisations to look again at means of cutting costs whilst at the same time 'leading corporations' are becoming increasingly aware of the benefits of being associated with the Corporate Social Responsibility (CSR) and sustainability agendas.

2.1 Private and Public Sector Features

SDG (2001) reports that private sector organisations develop Travel Plans to secure planning permission, portray organisational values, address site transport problems and for internal organisational reasons (to cut costs for example). In a study of 25 private sector organisations with Travel Plans in place over the long-term, Roby (2009) notes that initially 17 implemented a Travel Plan to get planning permission, one due to a UK Government directive, four for congestion/access arrangements and three for environmental reasons. However, more recently these motivations have now diversified significantly and other motivations now include enhancing car parking capacity, Corporate Social Responsibility, recruitment and retention and business growth. In other words, she suggests that if Travel Plans are to survive in the corporate world then they need to evolve from a transport policy tool into a business management tool.

By contrast, SDG (2001) notes that public sector organisations (which in Travel Planning terms generally means government, hospitals and universities) tend to be required to develop Travel Plans to meet regulatory requirements, improve environmental image and to act as beacons for other organisations.

National and local government offices are probably the most 'standard' of the three, in that staff commuting by civil servants is the dominant transport impact on the local community, while visitor, supplier and business travel seem rather less important. Essentially, as the bodies responsible for implementing public policy, there is an expectation that government should lead on policy initiatives such as Travel Plans by example. Accordingly, the UK Government set a target for government departments to have plans in place for all their headquarters buildings by March 1999, and for all other key buildings by March 2000 following the publication of the 1998 White Paper *A New Deal for Transport: Better for Everyone* (DETR 1998), and had the results evaluated in 2001. This reports that 522 of 1,576 buildings (around a third) had submitted Travel Plans and offers several interesting insights as to how the process went (NUTRI et al 2001). Specifically it reveals that:

- Travel Plans are not an organisational priority, judged by levels of management support, and by levels of funding for plans.
- Travel Plans have been developed primarily because there was a requirement to do so.
- Most Travel Plans consist of awareness raising and promotional measures, plus cycle facilities, and some attempts to limit business travel. In many cases these measures have been consciously chosen as 'achievable' and hence have not provoked great reaction from staff, either positive or negative.
- In consequence, most Travel Plan co-ordinators are sceptical that their plans will have anything more than a marginal effect on staff travel to work, although some have been able to effect changes in travel at work.
- Many of the plans studied had been implemented as far as possible in an initial flurry of activity at the time that Departments were required to write plans, and there is little commitment to its continued implementation through the organisation.

Hospitals and healthcare facilities have also long been involved in Travel Plans. Newson and Wenban-Smith (1998) suggests unique features of health facility sites are the requirements to meet the needs of patients, visitors and staff whilst maintaining access for emergency vehicles. In the UK context, DfT (2009) reports that healthcare facilities are the least accessible of key services by non-car modes, a position reflected by the fact that 83 per cent of trips associated with the UK National Health Service (NHS) are made by car (SDC 2007), and such site-access related issues more directly spill over into highly charged political debates over charging for parking spaces, adding to local congestion and negatively impacting on local neighbourhoods. Despite these pressures though, Griffiths (2006: 609–611) writes that:

> even in the largest employer and biggest spender in England [the NHS], only a relatively few Trusts have integrated sustainable development strategies … and that only a minority, mainly Acute [Hospital] Trusts, have made comprehensive environmental improvements in energy, waste, transport, procurement, and biodiversity, usually with thorough implementation in only one or two of these areas.

It then goes on to note that the environmental agenda is not a central priority for the Department of Health [which oversees the NHS], chief executives or board members of Trusts; and that there is a lack of joined up thinking between the two key groups of public health staff and estates and facilities managers. Finally, Griffiths suggests that the concept of corporate social responsibility has "so far taken stronger root in commercial companies than in the NHS". Perhaps as a response to such a situation, NHS hospitals are now being pushed to become 'good corporate citizens' by reducing their carbon footprint by 10 per cent by 2010, 26 per cent by 2020 and 50 per cent by 2050 (Pencheon et al 2009); and to take a lead in promoting active travel as a means of improving people's health (Wanless et al 2007).

Miralles-Guasch and Domene (2010) states that universities, the third public sector type to have taken up Travel Plans on a significant scale, represent a wide range of the population from different socio-economic backgrounds and ages that exhibit irregular schedules leading to a constant movement of people throughout the day. It adds that, particularly in suburban university campus environments, universities see daily commuting of the university population, long distances travelled and the predominance of private car use over non-motorised means of transport. On motivations, Balsas (2003: 46) concludes that "universities have the possibility to take a leadership role and promote environmentally sound programmes" thanks to their roles in developing research, teaching and service learning. Weenen (2000) too, reports sustainability becoming an increasingly important strategic priority of universities. This is further confirmed in Selby et al (2009), which suggests that the motivations for UK universities adopting Travel Plans were often originally spawned by the need to secure planning permission, but that their provenance has subsequently been ascribed to the green agenda.

Interestingly, in Britain (and in contrast with the NHS, at least where Travel Plans are concerned), such a situation has been partly driven by a proliferation of league tables coupled with a progressive commercialisation of the sector whereby universities are increasingly competing with each other in an ever widening number of areas, ranging from issues around research, teaching and student experience and into corporate social responsibility and environment. Consequently, the level of Travel Plan take up in universities is relatively high compared to other fields, with Selby et al (2009) noting that as of March 2008 some 64 of the 132 universities reviewed had a locatable and active Travel Plan; while ten were developing a Travel Plan, two were implementing a not yet active plan; four were re-developing a Travel Plan having withdrawn an earlier plan; and 52 either had no plan or had no locatable plan. In terms of trips targeted meanwhile, all 64 active University Travel Plans tackled commuter journeys, 42 considered business travel, 35 acted on visitor trips, 27 enforced fleet vehicle measures and 16 dealt with deliveries and contractor trips. No mention is made of what impact these plans have delivered. Furthermore, at this point it is also important to state that British universities will, from 2013, be required by the Higher Education Funding Council for England (HEFCE – a key funding provider in the UK) to include factors including type and level of business and commuting travel in their calculations of carbon for a carbon management scheme, and hence such competitive instincts are likely only to be intensified as this deadline approaches and beyond (HEFCE 2009).

2.2 Workplace Travel Plan Performance

Overall on the performance of Workplace Travel Plans, Rye (1999) notes that the main factors include:

- Firm size – larger organisations can access greater resources.
- Location – will influence type of measures adopted.
- Working patterns and activities – will influence type of measures adopted.
- Socio-economic status of the workforce – generally easier to influence blue-collar workers.
- Organisational culture and traditions – strong influence on policies that require changes in work practices such as working from home, compressed working weeks.

In practice, Dill and Wardell (2007) notes that financial incentives positively influence public transport use in Portland, Oregon. It adds that the success of a TDM programme is also dependent on human resources-related incentives (which could be flex time, compressed work weeks, company cars) and location factors such as street connectivity and public transport accessibility. And Rye (1998) draws on a survey sent to 1000 employers which had a response rate of 15 per cent. Nineteen percent of these were located in rural areas, and the average firm had 364 employees of whom 66 per cent travelled to work by car. Overall, free parking is provided by most organisations (only three charge for parking; more than half provide cycle facilities and 10 per cent provide works buses, while 15 per cent had attempted to influence the travel behaviour of their employees.

3 School Travel Plans

Schools are the second destination type where Travel Plans have been introduced. Here, the motivations are almost exclusively external – imposed by local or regional government for reasons of public policy (congestion reduction and environment), although Mingardo et al (2008) finds that concerns over children's health and road safety are seen as being more effective at convincing schools to adopt a Travel Plan as opposed to climate change and traffic congestion which are usually not mentioned. Finally, as with UK universities, inter-organisational competition too has been a driver in certain specific locations for schools, such as Györ in Hungary (Szakonyi, P. 2011. Personal communication, Transport Planning Officer, Municipality of Gyor, Hungary, Amsterdam, 25 February) and in London (Sheridan 2008) with each school wanting to prove itself as being 'the best'.

Once again the trips (particularly in the morning) occur in the peak, but here a second major issue is that of the escort journey for the majority of the primary school-aged pupils and a significant proportion of secondary school trips. As a result, and as alluded to earlier, parents rather than pupils tend to be the dominant decision maker in deciding how children get to and from school (Van Ristell 2011). Other School Travel Plan-specific design issues include the relatively small size of the organisations; the lack of any 'transport-skilled' personnel in school environments; and the lack of financial resources within the schools for School Travel Plans.

However, in spite of such handicaps, for some reason School Travel Plans are now commonplace in many countries around the world, and many of those that are in place have been relatively effective in delivering transport and wider objectives. So, for example, Hinckson et al (2009) reports the results of School Travel Plans for 35,153 students at 52 primary schools, seven intermediate and nine high schools in the greater Auckland region in New Zealand, and notes that the modal shift of walk and cycling increased by around 2.5 per cent, public transport rose by 1 per cent, and car use fell by around 3.5 per cent. It adds that students from low income backgrounds were more likely to shift modes and that the greatest improvements were made in the first year with improvements becoming progressively smaller.

In analysing why this is the case, such a positive position is probably due to three main reasons.

First, is the presence of a (confluence of) high profile political issues – worsening congestion generated by the 'school run', increasing childhood obesity and corresponding concerns about a decline in the independence of young people, all apparently due to increased car use for education journeys.

Second, School Travel Plan officers have proved rather innovative in the measures adopted to encourage children and their parents to cut down on their use of the car for the school journey. Hence, walking buses and cycle trains (where volunteer parents escort groups of pupils along a route from homes to school) provide a straightforward, 'off-the-shelf', low cost, readily transferrable, effective and incentive based measure for any school anywhere to adopt. Similarly a whole range of competition packs, where pupils win prizes for travelling 'green', and classroom materials through which 'sustainable travel' messages are conveyed through mathematics, environmental awareness and geography lessons have been developed that capitalise on the competitive instincts and 'pester power' of the young people to wear down parental objections to not using the car (Sheridan 2008).

Third, and perhaps most crucially, is the commitment of significant financial, personnel, political and policy resources by national and local government and the school communities themselves to making School Travel Plans work, although sadly it remains to be seen whether such commitment can or will remain in the face of global economic recession and the associated reductions in public financial support likely over the next few years.

4 Event-focused Travel Plans

This section focuses on Travel Plans developed for planned events such as sporting fixtures, concerts, festivals and road works, as opposed to unplanned events such as emergencies caused by extreme weather or other major catastrophes. Thus, "a planned special event is a public activity with a scheduled time, location and duration that may impact the normal operation of the surface transportation system due to increased travel demand and/or reduced capacity attributed to event staging" (Dunn 2007: 1).

In categorising these, Dunn (op cit) presents a useful table detailing five types of event, their characteristics and their community impacts (see Table 3.3).

Table 3.3 Planned events, their characteristics and impacts

Category	Characteristics	Effect on community
Discrete/ recurring event at a permanent venue.	Predictable start and end times. Known venue capacity. Advance ticket sales. Concentrated arrival and departure demands.	Need to give special attention to weekday events due to their potential impacts on commuter traffic, transit operations, business and residents. Need to minimise impacts on access to businesses and residents due to concentrated arrival and departure demands. Need to identify ways to minimise community impacts and improve transportation system operations during future venue events. Need to consider parking restrictions or issuance of residence parking stickers so that patrons do not impact residences by parking on residential streets in the vicinity of the venue.
Continuous event	Occurrence often over multiple days. Patrons arrive and depart during the event day. Less reliance on advance ticket sales. Capacity of venue not always known. Occurrence sometimes at temporary venues. Parking availability varies.	Need to ensure that the significant level of personnel required over the multiple days of operation can be provided to minimise impacts to the community. Need to apply changes made in the traffic management plan to minimise community impacts for future continuous events.

Table 3.3 ***Continued***

Category	Characteristics	Effect on community
Street use event	Occurrence on roadway requiring closure. Specific starting and predictable ending times. Capacity of spectator viewing area not known. Spectators typically not charged or ticketed. Parking availability varies. Impact on emergency access and local services.	Need to consider effect on access and parking of residents and businesses of a street along which the event will be held. Need to consider communicating with and involving the participation of businesses and residents on the street that will be closed for the event. Need to assure that bus transit stop relocations during the event are communicated in advance to residents and businesses. Need to require adequate sidewalk passage area and fire lanes at all times to permit access of emergency vehicles to adjacent residents and businesses along the streets utilised by the street use event. Need to notify all businesses and residents located on the street to be used or closed of the event dates and times.
Regional/multi-venue event	Occurrence of events at multiple venues at or near the same time. Ingress and egress operations for concurrent events may occur at the same time. Parking areas may service demand from different events over period.	Need to coordinate activities of the events in order to minimise traffic impact on the community. Need to offset occurrence of ingress and egress times to minimise effect on traffic, parking, resources and the community. Need to examine event scheduling to avoid departing traffic from one event coinciding with another event to minimise effect on traffic, parking, resources and the community.
Rural event	Rural area and possible tourist destination. High attendance events attracting event patrons from a regional area. Limited roadway capacity. Area lacking regular transit service.	Need to meet with residents and businesses to minimise effect on adjacent roadways. Need to use public surveys to assess all community effect and incorporate successful lessons learned to apply to the next event.

Source: Dunn (2007)

From this table, a number of examples can be examined to illustrate some of these categories in a little more detail.

For instance, Grava and Nangle (2000) finds that conditions affecting arrival patterns to baseball stadia (a regular event at a permanent stadium) include: proximity to the city centre, availability of public transport facilities and the vagaries of the playing season. It also finds that good highway and public transport access produces sharp arrival peaks before the game (requiring efficient parking and stadium access arrangements). However, whilst arrival activity affects transport planners, for facility managers it is the departure activity that is most crucial. Specifically, stadia need to be able to release spectators into the surrounding transport system without overloading it, whilst ensuring that the majority are clear within an hour at most.

On similar lines, faith sites too have specific characteristics. Thus, Goodship (2008) points out that 'facilitating factors' include a moral imperative to act, local catchments, regular trips and the influential role of the worship leader; while the 'obstructing factors' are that Faith Travel Plans would need additional resources compared with workplaces, the lack of records in some cases of where worshippers come from, language issues and possible resentment by religious organisations of 'outside interference'.

Finally, the Olympic Games provide an extreme example of a regional/multi-venue event. Here, Currie and Delbosc (2011) finds that transport authorities must operate an entirely new transport system over a two week period for over 40,000 athletes in addition to catering to many millions of additional spectator trips in cities that often experience significant urban 'base load' of everyday journeys on already congested and stressed transport systems where capacity is stretched. It continues that to accommodate these trips an approach is to 'under promise' and 'over deliver' such that as many people as possible are persuaded to travel only when absolutely necessary around the venue. However, there is also an opportunity here whereby people will try non-car modes perhaps for one of the few times in their life and may therefore be open to changing their behaviour.

One event type not referred to in Dunn's framework is large road works, and this has actually proved to be one of the more widespread and successful forms of Event Travel Plan intervention. Thus, large scale road works on major corridors in the Netherlands have been used as a trigger for persuading organisations and individuals to participate in mobility management schemes (de Munck et al 2008). Here, in addition to the highway authorities programming maintenance during off-peak periods (evenings and weekends) and providing advanced warning to motorists, consultants were used to engage employers in some of the most directly affected sites. Experience has largely been positive. Between May and September 2005 major road works on the A9 Gaasperdammerweg Motorway led to employers jointly funding a mobility broker, dedicated travel passes, free shuttle buses to stations and personal travel advice. As a consequence, car use fell from a mode share of 58 per cent before, to 40 per cent during the road works and 55 per cent afterwards. Public transport use meanwhile rose from 29 per cent before, to 47 per cent during and 31 per cent afterwards. Similarly the A4–A10 road works during 2006 led to reductions in traffic volumes of 15 per cent during the morning peak and 11 per cent in the afternoon, of which half were attributed to mobility management measures. Here, though, there was no significant structural change in travel behaviour as such. Nevertheless, as a consequence there is now an expectation in the Netherlands that all large scale road works schemes will be supported by some form of complementary Travel Planning process.

One final point to note here is that, as with the Olympic Games experience above, road works again provide a trigger for persuading people to try alternative solutions to using the car, a potentially valuable opportunity for changing attitudes and perhaps ultimately behaviour.

5 Travel Plans to Attractions

From an issue perspective, Newson et al (2001: 4) states that car traffic "can cause particular problems for tourist areas, damaging the beauty of natural landscapes, choking the narrow streets of honey-pot towns and villages and intimidating visitors who wish to walk and cycle. Ultimately it can threaten the very qualities that attract tourists in the first place". Meanwhile Dickinson and Robbins (2008: 1112) look at the characteristics of tourism activities, and finds that:

> especially in rural areas, [they] tend to be very dependent on the car. Attractions are generally dispersed and isolated from public transport routes, modern leisure pursuits can require the transportation of bulky items, traffic is generated at varied times, the journeys are not part of a daily routine and visitors are unfamiliar with the area and public transport infrastructure.

Loosely defined, attractions with Travel Plans include museums and areas of natural beauty. Guiver et al (2008: 145) reports the UK position that while some tourist organisations have adopted a site-based approach to Travel Planning (for example, Youth Hostel Association, the National Trust and one off examples such as Harewood House and Chatsworth House), in general "tourist attractions have been slow to embrace the concept", despite their being promoted in UK Government policy as long ago as 1999 in the White Paper *Tomorrow's tourism* (DfMCS 1999).

Guiver's paper then reports on a study of the opinions of site managers of the Roman-built monument Hadrian's Wall in the north of England, and finds that they stress a lack of influence of how visitors access the site. Instead, visitors decide how to travel based on the weather, foreign exchange rates, competing leisure activities and fashions. Moreover, most choose their mode of travel before the destination meaning that improvements for non-car travel were not seen to affect the number of people travelling by car. Finally, the managers see travel to sites as being peripheral to the core mission – which is to encourage more visitors, enthuse people about Roman history, preserve the World Heritage Site, and generate visitor spending.

Looking to National Parks, Regenerus et al (2007) looks at traffic management in the Veluwezoom National Park in the Netherlands, and finds implementing travel measures to be difficult due to a lack of knowledge about the recreational use of the area coupled with the interdependence of the various actors involved in the process (each seeking to meet their own, sometimes contradictory, objectives).

Similarly, Kendal (2011) details a catalogue of abandoned forays into dealing with transport issues in the UK National Park context using TDM approaches but notes that on the other hand, efforts in the USA appear to have been more successful.

Overall Guiver et al (2008: 144) comments that "attempts to reduce car use in tourist areas have concentrated on the use of incentives … rather than disincentives … , because the latter are considered to be much less acceptable to visitors and residents".

6 Retail Travel Plans

As indicated previously, shopping trips can be categorised as being convenience or comparison (that is of applying to food or non-food outlets), and Retail Travel Plans can be thought of in the same way. Interestingly, unlike for other sectors there appears to be little specialist government guidance available beyond that provided in generic Employer Travel Plan information. Despite this, there are some good examples of Retail Travel Plans extant, particularly in the UK although actually food retailers (for example, Tesco, Sainsburys, Asda, Waitrose, Morrisons) have not tended to promote their involvement in any Travel Plan measure other than home delivery beyond occasional locally-targeted press releases. Also, Retail Travel Plans in general seem to focus more on changing the behaviour of staff than customers.

One well known early (non-food retailing) adopter of the Travel Plan concept was the Body Shop cosmetics chain, which was motivated predominantly by the ethical dimension of reducing car use to its sites. 'Do-it-yourself' chain B&Q is also recognised as a 'good practice' Travel Plan organisation and has had a policy of rolling out Travel Plans at new developments since 2002, again as a means of being seen as a 'good neighbour'. Furniture seller IKEA provided each of its 9,000 UK staff with a folding bike in December 2006 to encourage modal shift as part of its Travel Plan, along with public transport tickets discounted by 15 per cent. And department store chain Marks and Spencer includes several transport measures in the climate change element of its *How we do business* strategy (Marks and Spencer 2010), including:

- progressively introducing facilities to encourage the use of public transport and cycling for both customers and employees where appropriate;
- improving logistical efficiency by working with suppliers to understand how we can improve the efficiency of deliveries to distribution centres, including increasing the use of rail transport and reducing air freight through the re-tendering process;
- trialling a green vehicle fleet;

Finally, several large regional shopping centres in the UK including Milton Keynes, the Metro Centre in Gateshead, Meadowhall, near Sheffield and

Bluewater in Kent have also taken up Travel Plans. As with the previous examples in general this was to meet planning requirements, but subsequently Corporate Social Responsibility and environmental management reasons have become important drivers.

Relating to measures, up to now the emphasis appears to have been on increasing levels of trip substitution through home shopping (especially for food shops); increasing the efficiency of supplier deliveries; and of providing enhanced information about non-car alternatives. In addition, improvements to public transport services are occasionally offered. However, perhaps the most significant issue for the large number of supermarkets located away from urban and district centres has largely been ignored – that of free parking. Fundamentally, retail outlets perceive that they will lose custom to competitors if they charge for parking and so are unlikely to act unilaterally to change the situation. And so far, despite occasional proposals threatening to support town centre stores by taxing edge and out-of-town parking spaces, politicians have not yet gone ahead and tried to implement such a programme. Yet until this happens, it is difficult to see Retail Travel Plans becoming more than a form of 'eco-wash' treatment of the issue.

7 Travel Plans for Transport Interchanges

Unlike the other segments discussed, Travel Plans or 'access strategies' for transport interchanges could be perceived as being (theoretically at least), part of the core business of an airport or a rail station for example. That said, for airports with over a million passengers per annum a primary driver for so-called 'surface access strategies' and the 'surface access fora' to be adopted stems from a legal requirement in the 2000 Transport Act (UK Parliament 2000). Nevertheless, the evidence suggests that several of the largest airports at least have seriously engaged with the Travel Plan concept.

Thus, experience at Stansted Airport, currently the third largest UK airport serving around 20 million passengers a year has seen the modal share of single occupancy car drivers fall from 94 per cent (of 6,800 staff) in 1997 to 71 per cent (of 11,000 staff) in 2009 (Gregory 2010). In terms of measures, a Commuter Centre was opened in 2007 to provide staff with travel information and a range of incentives such as heavily discounted public transport passes; a preferential parking area is in place for car sharers; and the airport is one of the few areas in the UK to operate a public transport network 24 hours a day; while these incentives are financed through levies on car parking charges (£GBP12 (€11) per year of staff parking charges and 30p (€0.27) per passenger are earmarked to pay for enhanced alternatives to the car). Crucially, Travel Planning at Stansted is also integrated with wider business objectives including cost saving strategies and Corporate Social Responsibility and environmental accreditation procedures. Similarly, significant Travel Plan efforts have been made at Heathrow Airport, Gatwick, Manchester and East Midlands Airport.

However, the picture is not always so positive. Former surface access manager at Birmingham Airport, Stephen Holt, notes that airports away from very large cities like London face bigger challenges in influencing car use due to much lower use by non-UK nationals, the lack of a significant corridor flow into a central location, inadequate public transport networks and a lack of a public transport using culture within the catchment area (Holt 2010). He then adds that two often made assumptions relating to surface access to airports are actually myths. Specifically that airport growth will generate a critical mass making public transport suddenly viable, and that growing airports have lots of money to spend on expensive transport schemes. Finally, he notes that there are several key reasons why people drive to airports that are difficult to challenge. So, for example air passengers tend to need more luggage than for other trip types; do the trip infrequently and so have poor knowledge of alternatives; have a high desire for reliability (public transport is often perceived as being risky); while the flight is generally fairly expensive meaning that the high parking charges are not a deterrent. Moreover, higher parking charges may often stimulate more trips to be made – whereby passengers have a family member or friend drop them off and pick them up at the terminal instead (which is known as kiss-and-ride). Interestingly, though this is a more difficult activity to control, there are now examples (such as Edinburgh Airport) where drivers are now charged for entering the terminal in a bid to do just that. Meanwhile, airport staff are difficult to target due to the single use activity levels of "business parks with planes" being insufficient to justify comprehensive public transport services; services in unsocial hours being rarely viable; green field sites resulting in long walking and cycling trips; and airport operators being reluctant to charge for car parking due to that often being the major attraction for people working at an airport.

Looking to rail interchanges, although parking charges have been commonplace for many years and one or two stations (Bicester North, near Oxford in the south of England for example) have even provided 'feeder buses' as a means of dealing with parking stress, the formal emergence of Station Travel Plans in the UK only occurred in 2007, following the publication of *Delivering a sustainable railway* (DfT 2007). This led to an announcement that 31 stations had been selected to be funded by the UK Government in 2008 in a bid to bring together stakeholders with an interest in improving access to rail stations (ATOC 2011). Subsequently, several other UK stations have also adopted the idea, for example in London, Essex, Warwickshire, and Bedfordshire.

Such schemes involve appointing a Travel Plan coordinator, passenger and stakeholder surveys and consultations, improved information provision, infrastructure (for example, cycle parking spaces, modified road junction access points) and public transport service improvements. As of 2010, preliminary results at the half way stage suggest 29 of the 31 stations (23 of 24 pilot bids, which were made by local authorities) remain engaged and that these have all begun to implement measures though as yet the 'after' results have not been collected.

The main problem appears to be that the programme allowed insufficient time to implement Travel Plans at each site (SDG 2010).

Overall, Travel Plans to transport interchanges are likely a phenomenon found mainly in the UK, where they have seemingly emerged as a response to the rather fragmented network of transport policy actors (abroad, the concept of intermodal integration usually seems to happen almost naturally, as a key part of a local area's transport plan). Even more strangely, one of the more effective car restraint policies – the parking charge – has often become so not because of any decision to limit car traffic for reasons of on-site congestion relief or to allow valuable land to be better used for other uses (as for other Travel Plan types), but instead as a means to maximise revenue. Indeed, parking charges, along with retail outlets, are often the only areas where the transport infrastructure operator can do this that are not regulated. That this approach can be adopted at airports and rail stations, and not elsewhere at retail outlets, visitor attractions, or hospitals (at least not to the same extent), is because the users of transport interchanges are often restricted in their choice of destination, while the car parking element cost of their trips is still generally far less than their air or rail ticket. In addition, clearly the interchanges able to make higher charges are those where the demand for the parking space and the air or rail trip are also high.

Given these situations then, it is interesting that while the Travel Plans implemented at many transport interchanges may have been required by governmental regulations initially, the most fundamental elements of these are now also driven by a strong commercial rationale which sees operational decisions being fully integrated into wider business processes, at least at sites where capacity is felt to be constrained.

Lastly, at this point, it is pertinent to mention the US concept of Transit Oriented Development (TOD) – whereby higher density developments with reduced parking standards are encouraged in the vicinity of rail stations – which can therefore perhaps be seen as being a rather longer term and more strategic approach to Travel Planning at stations (see Curtis et al 2009 for more on the issues around TOD in practice).

8 Residential Travel Plans

By contrast with the Travel Plan types already discussed, the Residential Travel Plan this time focuses on trip origins rather than trip destinations.

Notionally the term Residential Travel Plan (RTP) first seems to have emerged in the UK in 2007 with *Making Residential Travel Plans Work: Guidelines for New Development*, (DfT, 2007), although the concept of low car and car free neighbourhoods from which RTPs are descended can be traced back to a series of car free developments in Germany (most notably in Hamburg, Bremen and Freiburg) and Austria in the early 1990s (see Morris et al 2009).

In identifying the scope for car free developments (and by extension RTPs), Melia et al (2011) reports that demand for car free housing is concentrated amongst 'car free choosers' – people who currently live without cars by choice. These are mainly found in the inner areas of larger cities, where the greatest potential for car free development exists although a substantial minority of them would like to move to less urban locations. The paper adds that proximity to good rail services would be a key requirement in these circumstances.

Transport 2000 Trust et al (2005) points out that Residential Travel Plans differ from destination-based plans in several ways. Firstly, in that the pattern of journeys served is more varied – residents having multiple destinations and different needs and travel choices over time. Secondly, there is a need to establish an ongoing management structure to deliver the Travel Plan given that there is generally no single company or organisation to provide a common focus for residents. In other words, there is a generally weak relationship between mediating/delivery agent (that is, the residential provider) and the trip makers (residents).

As a result, in the author's opinion, it is unlikely that RTPs will ever be as widespread or as effective as other forms of Travel Plans, apart from where specific circumstances exist – such as in cases where there are strong car restraint policies and excellent alternatives to the car in the surrounding area, coupled with the presence of a strong community body for the residents.

9 Personal Journey Planning

As (some of) the various terms (Individualised Marketing, Personal Journey Planning, Travel Blending, Travel Feedback Programmes) suggest, Personalised Journey Planning (PJP) is Travel Planning focused at the level of the individual.[1] On the one hand, this makes them an outlier in that they are Travel Plans tailored to the ultimate degree. On the other hand, by advancing to the individual level, the key concept of the mediating institution is effectively bypassed and thus it could be argued that PJPs are not strictly speaking Travel Plans at all, but instead are an informational TDM instrument.

Assuming the former case for the moment, PJPs typically involve either a transport authority and/or a partner organisation (employer, residential developer, school) meeting with employees/residents/pupils, who then discuss their current travel behaviour and the reasons behind the choices they make and subsequently devise potential 'greener' travel alternatives. These often comprise of individually tailored information about local public transport, walking and cycling, and perhaps

1 It should be noted that each term refers to a slightly different technique that varies as to whether or not it motivates travel behaviour change; whether it requests a plan for changing travel behaviour; and whether it provides individualised information. See Fujii and Taniguchi (2006) for further information.

some form of discounted tickets to encourage the target to try one or more of the alternatives suggested.

PJPs first came about in their current form in Germany in the mid-1990s and have since spread worldwide, most notably to Australia, Scandinavia, the UK and Japan.

As far as results are concerned, Fujii and Taniguchi (2006) reviews ten 'travel feedback programs' (TFPs) in Japan and reports reduced carbon emissions and car use of 19 per cent and 18 per cent respectively, and increased public transport use of about 50 per cent. Slightly less effective results are reported in the UK. Cairns et al (2004) states that PJP schemes in different parts of Bristol cut traffic by between 5 and 10 per cent, and a pilot project generated reductions of 11 per cent in car driver trips in London. However, such results are relatively expensive (though they do benefit from economies of scale), and so more recent efforts have focused on trying to deliver a similar level of benefit for a substantially reduced cost, most typically by adopting a less intensive approach.

10 Freight Travel Plans

Unlike the other types of Travel Plan covered so far, this category of Travel Plan does not focus on the movement of people but of freight. Consequently, due largely to the traditional division in transport between the passenger and freight sectors, it is an area that has largely been ignored by the Travel Plan industry. This is unfortunate because the logistics industry is in many ways more advanced and offers some interesting pointers for Travel Planners.

However, there is some evidence that this 'missing link' is being bridged in some locations. Thus, Cherrett and Triantafyllou (2009) reports that projects in London, Winchester and Bristol in the UK, Barcelona in Spain and several cities throughout Germany have set up partnerships to allow goods to be transported more efficiently through some form of consolidation centre approach. Meanwhile, Finnegan et al (2005) reports that establishing a consolidated urban distribution centre in Dublin eliminated 501 lorry-based deliveries to shops in Dublin each week, along with a reduction in backhauling. Instead, more than 1,000 items are now delivered by walk and cycle couriers. Moreover, there are no significant delivery time losses as reliability is improved because they are no longer affected by traffic congestion.

11 Segments Compared

The purpose of this chapter has been to examine Travel Plans on a segment by segment basis in order to ascertain specific issues in each case. At this point, it is therefore possible to summarise this information and to develop a framework highlighting the strengths and weaknesses of these alternative types of Travel Plan.

This summary has been conducted in two parts. First, Table 3.4 outlines the suitability of a range of trip types for targeting by Travel Planners. Second, Table 3.5 incorporates this information together with a range of other 'organisation-type' features to provide a rough assessment of the suitability of Travel Plans for a number of 'trip generators' (segments).

From Table 3.4, it can be seen that three types of trip seem extremely well suited to Travel Plan measures – business trips, supplier trips and emergency patient trips to hospital – whilst commuting and education trips (journeys to and from work and to and from school respectively) also score highly. This is predominantly due to the high degree of control exercised by the Travel Plan organisation over the individual. So, for example, emergency hospital patients tend to require ambulance-based transport, whilst business and supplier trips are generally paid for direct by the employer. Of course organisations also have some influence over work/school commuting through their operational practices relating to attendance, level of flexibility and so on, and usually have well established personal relationships (potentially allowing tailored solutions to be developed). In addition, such trip types also 'benefit' from occurring at peak times, and being frequent and regular, theoretically making it easier for less car dependent transport services to be provided and/or improved.

Less easily targeted are shoppers, though there is scope particularly where loyalty card schemes provide personal information on habits and if the 'problem' of what to do with the shopping can be resolved. However, it is visitors, leisure customers and non-emergency hospital patients who seemingly provide the greatest challenge to Travel Planners. Here, with a few exceptions, the Travel Plan organisation enjoys only limited leverage or none at all over the individuals, and the (proportionally small number) of trips are often unpredictable, infrequent, and irregular, yet are also longer distance and hence significant contributors to community level transport problems.

Applying the user and trip characteristics to Table 3.5 provides an initial indicator as to the most effective type of trip generator to be targeted. However, there are also several other factors to consider. These include organisational characteristics such as size, location, and type of site; the degree of control that government has over the Travel Plan organisation (through information, regulation or fiscal mechanisms for example); and the level of financial and transport planning resources available for the Travel Plan.

From the grid, it can be seen that the context for Travel Plans would seem to be most appropriate for (very large) public sector employers and for transport interchanges – though for quite different reasons. Thus, while both destinations are strongly influenced by government policy and benefit from economies of scale, the government offices, universities and hospitals score well with the types of trip served (though less well on the resource profiles), whilst the transport interchanges exhibit the opposite.

Table 3.4 Trip types and their suitability for Travel Planning

User type	Trip type	Trip characteristics						Degree of control	TP context
		Freq.	**Peak?**	**Indiv./ group**	**Freight/ luggage**	**Reliability**	**Cost constraint**	**Org. on indiv.**	
Staff	Commute	High	Peak	Indiv.	Not often	Important	Medium	Medium	High
	Business	Varies	No peak	Indiv.	Not often	V. important	Low	High	V. high
Visitor	Business	Varies	No peak	Indiv.	Not often	Important	Low	Low	Low
	Pers. Bus.	Varies	No peak	Indiv.	Not often	Varies	Varies	Low	Low
	Leisure	Varies	No peak	Group	Yes	Less important	High	Low	Low
Supplier	Business	Varies	No peak	Indiv.	Yes	Important	Low	High	V. high
Customer	Shop	Varies	Peak	Indiv.	Usually	Less important	High	Medium	Medium
	Leisure	Varies	Peak	Mixed	Not often	Important	Medium	Medium	Low
Student (School)	Educ.	High	Peak	Escort/ Indiv.	Not often	Important	Medium	Medium	High
Student (Univ.)	Educ.	High	Peak	Indiv.	Not often	Important	High	Medium	High
Patient (Emergency)	Health	V. low	No Peak	Escort	No	V. important	V. low	High	V. high
Patient (Non-Emergency)	Health	Varies	No Peak	Mixed	No	Important	Varies	Medium	Low

Table 3.5 Trip generators and their suitability for Travel Planning

Trip generator		User types	Trip types	Organisation characteristics		Degree of control	Resources		TP context
	Examples			Size	Site	Gov't-org.	Finance	Tspt.	
Employer (Private sector)		Staff, Supplier, Visitor, Customer	Commute, Business,	V. small – v. large	Single/ multi-occ.	Limited	Limited	Limited	Strong
Employer (Public sector)	Gov't, University, Hospital	Staff, Supplier, Visitor	Commute, Business, (Educ., Health)	Medium – v. large	Single/ multi-occ.	High	Low	Limited	Very strong
School	Primary, Secondary	Student, Staff, Supplier, Visitor	Educ., Commute, Business	Medium	Single occ.	High	Low	Low	Strong
Events	Sport, Concerts, Festivals, Road works	Visitor, Customer, Staff, Supplier	Leisure, Commute, Business	V. small – v. large	N/A	Limited	Limited	Limited	Strong
Attractions	Faith, Museums, National Parks	Visitor, Customer, Staff, Supplier	Leisure/ Pers. Bus., Business.	V. small – v. large	Single/ multi-occ.	Limited	Limited	Limited	Weak
Retail	Food/ Non-food	Customer, Staff, Supplier	Shop, Commute	V. small – large	Single/ multi-occ.	Limited	Limited	Limited	Weak
Transport	Airport, Rail Stations	All of the above	Business, Leisure, Commute	Large – v. large	Single/ multi-occ.	High	Medium	High	Very strong
Residential		Residents	All of the above	V. small – medium	Multi-occ.	V. Limited	Limited	Limited	Very weak

Looking next at a second level of suitability for Travel Plans, private sector employers benefit from strong user and trip profiles and from economies of scale but are generally not highly influenced by government with regard to Travel Plans, unless:

- they require planning permission or need to meet environmental or transport-related regulations;
- fiscal incentives are provided that make Travel Plans attractive to employers;
- Travel Plan measures are made attractive to employers in other ways.

Meanwhile schools can be highly influenced by government agencies and have suitable user and trip characteristics but are generally not particularly large trip generating sites nor have any financial or transport resources to apply. And, Travel Plans have also proved quite effective at (large scale) planned events, particularly in cases where the threat of serious traffic and/or parking problems coupled with comprehensive information about plausible alternatives to car travel have been communicated to potential attendees beforehand.

Less suitable still would be attractions and retail segments due to apparently unfavourable characteristics almost across the board – often unknown users, irregular and infrequent trips – perhaps with 'luggage' – often edge of town or rural locations, limited level of government influence and limited resources.

Lastly, it would seem that the segment least likely to succeed would be the Residential Travel Plan. Here, the difficulties centre on the generally very small size of the individual 'family' units, the lack of control exerted by a mediating organisation over residents and the limited resources, but perhaps most damagingly on the sheer diversity of the range of trip types and users served.

However, as already noted, there are some exceptions. For example, the Eden Project in Cornwall in south west England is an attraction located in a rural area (and so in theory should not be particularly suitable for Travel Plan), but its large-size coupled with its function as a catalyst for environmental change has seen a great deal of effort applied and a relatively effective Travel Plan has emerged. Similarly, Residential Travel Plans can also work if the site is located in a 'sensible' place served by good enough alternatives to the car; if owning a car is difficult due to a lack of available parking; and/or if there is a strong community network in place. Conversely, there are a myriad of examples whereby public sector employers for instance, have not set up a Travel Plan in spite of the apparent advantageous Travel Plan context.

And, there is also some sense in considering Travel Plans in certain circumstances as a means of educating particular groups about the availability of alternatives to the car. So, School Travel Plans help influence the attitudes of children in how they see the car for example, whilst Event Travel Plans (for anything from the Olympic Games to long-term road works) have been shown to 'persuade' or trigger car users to try other modes due to special circumstances which they probably otherwise would not have done.

Interestingly, one other theme that consistently appears to inform the success or otherwise of Travel Plans across the various segments examined is that of the size of the trip generators/attractors, and it is this relationship that will be explored in Chapter 4.

Business travel has tended to be relatively ignored by Travel Planners, in spite of its apparent suitability for treatment

Railway company Chiltern Railways operates demand responsive feeder buses to Bicester North Station in Oxfordshire to relieve the pressure on car parking and so enhance revenues

References

Alamgir, F. 2010. *Improving Business Travel within Organisations through the Application of Information Management Systems*, Unpublished First Year PhD Report, (Loughborough: Loughborough University).

Anable, J. 2005. 'Complacent car addicts' or 'aspiring environmentalists'? Identifying travel behaviour segments using attitude theory, *Transport Policy*, 12(1), 65–78.

Association for Train Operating Companies. 2011. *ATOC Station Travel Plans: National Pilot Programme.* [Online] Available at http://www.stationtravelplans.com [accessed 6 April 2011].

Beaverstock, J. V., Derudder, B., Faulconbridge, J. and Witlox, F. 2010. *International Business Travel in the Global Economy*, (London: Ashgate).

Balcombe, R., Mackett, R., Paulley, N., Preston, J., Shires, J., Titheridge, T., Wardman, M. and White, P. 2004. *The Demand for Public Transport: A Practical Guide*, TRL Report 593, (Crowthorne: Transport Research Laboratory).

Balsas, C. J. L. 2003. Sustainable transportation planning on college campuses, *Transport Policy*, 10, 35–49.

Cairns, S. 1997. *Shopping and Personal Business Trips*. Paper presented to a RAC and Transport 2000 Seminar on Car Dependence, London, 17 November.

Cairns, S. 2005. Delivering supermarket shopping: more or less traffic? *Transport Reviews*, 25(1), 51–84.

Cherrett, T. and Triantafyllou, M. 2009. *Travel Planning for Freight*. Paper presented at the Lunch and Lecture at Loughborough, Session 6, Loughborough University, Leicestershire, 8 May.

Currie, G. and Delbosc, A. 2011. *Travel Demand Management for the Summer Olympic Games 3: Approaches and Impacts*. Paper presented to the 90th Annual Meeting, Transportation Research Board, Paper 11-1138, Session 231, New Transportation Demand Management approaches to influence mode choice, Washington DC, 23–27 January.

Curtis, C., Renne, J. L. and Bertolini, L. 2009. *Transit Oriented Development: Making it Happen*, (Aldershot: Ashgate).

De Munck, G. B. G., van der Linden, L. H. K. and de Baar, A. H. J. 2008. Managing mobility during road works. Paper presented to the *European Conference on Mobility Management*, London, 6–8 June.

Department of the Environment, Transport and the Regions (Great Britain). 1998. *A New Deal for Transport: Better for Everyone* (London: The Stationery Office).

Department for Media, Culture and Sport. 1999. *Tomorrow's Tourism*, (London: The Stationery Office).

Department for Transport (Great Britain). 2007. *Delivering a Sustainable Railway*, (London: Department for Transport).

Department for Transport (Great Britain). 2009. *Transport Trends*, (London: Department for Transport).

Department for Transport (Great Britain). 2010. *Transport Statistics Great Britain*, (London: Department for Transport).

Dickinson, J. E. and Robbins, D. 2008. Representations of tourism transport problems in a rural destination, *Tourism Management*, 29, 1110–1121.

Dill, J. and Wardell, E. 2007. Factors affecting worksite mode choice: findings from Portland, Oregon, *Transportation Research Record,* 1994, 51–57.

Dunn, W. 2007. *Managing Travel for Planned Special Events Handbook: Executive Summary*, Project, DTFH61-01-C-00180, (Washington DC: Federal Highway Administration), June.

Finnegan, C., Finlay, H., O'Mahony, M. and O'Sullivan, D. 2005. Urban freight in Dublin city centre, Ireland: survey analysis and strategy evaluation, *Transportation Research Record*, 1906, 33–41.

Fujii, S. and Taniguchi, A. 2006. Determinants of the effectiveness of travel feedback programs – a review of communicative mobility management measures for changing travel behaviour in Japan, *Transport Policy*, 13, 339–348.

Global Expense. 2008. *Employee Expenses Benchmarking Study Report 2008*, (London: Global Expenses).

Goodship, A. 2008. *Travel Planning for Faith Destinations*, Unpublished MSc Project, (Loughborough: Loughborough University).

Grava, S. and Nangle, F. 2000. Get me to the ball game on time: access time patterns at Baseball Stadia, *Transportation Research Record*, 1735, 91–97.

Gregory, J. 2010. Stansted airport surface access strategy. Presentation to TRAVEL PLAN PLUS Partners, Stansted Airport, Essex, 28 November.

Griffiths, S. 2006. Environmental sustainability in the National Health Service in England, *Public Health*, 120, 609–612.

Guiver, J., Lumsden, L. and Weston, R. 2008. Traffic reduction at visitor attractions: the case of Hadrian's Wall, *Journal of Transport Geography*, 16, 142–150.

Higher Education Funding Council for England. 2009. *Sustainable Development in Higher Education: 2008 Update to Strategic Statement and Action Plan*. Policy development: Statement of policy 2009/03, (London: HEFCE), February. [Online] Available at http://www.hefce.ac.uk [accessed 5 April 2011].

Hinckson, E., Duncan, S. and Badland, H. 2009. Can school travel plans change the way children commute? *Journal of Science and Medicine in Sport – Abstracts*, 12S, S10.

Holt, S. 2010. Surface access at regional airports: understanding the real issues. Paper presented at the *Lunch and Lecture at Loughborough*, Session 12, Loughborough University, Leicestershire, 3 December.

Kendal, J. J. 2011. *Investigating the Process of Policy Change with Respect to Leisure Transport in UK National Parks*. Unpublished PhD Thesis, (Loughborough: Loughborough University).

Marks and Spencer. 2010. *How we do Business Report 2010*, (London: Marks and Spencer).

Melia, S., Barton, H. and Parkhurst, G. 2011. Potential for carfree development in the UK, *Urban Design and Planning*, 164, in press.

Miralles-Guasch, C. and Domene, E. 2010. Sustainable transport challenges in a suburban university: the case of the autonomous University of Barcelona, *Transport Policy*, 17(6), 454–463.

Mingardo, G., Berg, L. and Haaren, J. 2008. Transport, environment and economy at the urban level: the need for de-coupling – a comparative study of four European cities: Rotterdam, London, Gothenburg and Hamburg, (Rotterdam: EURICUR).

Morris, D., Enoch, M. P., Pitfield, D. E. and Ison, S. G. 2009. Car-free development through UK community travel plans, *Urban Design and Planning: Proceedings of the Institution of Civil Engineers*, 162(DP1), 19–27.

Morrison, A. M., Ladig, K. A. and Hseih, S. 1994. Corporate travel in the USA: characteristics of managers and departments, *Tourism Management*, 15(3), 177–184.

Napier University Transport Research Institute, Open University and W. S. Atkins. 2001. *Evaluation of Government Departments' Travel Plans*, (London: Department for the Environment, Transport and the Regions).

National Transport Authority (Republic of Ireland). 2010. 2010 smarter travel workplaces, National Transport Authority, Dublin, Ireland. [Online] Available at http://www.smartertravelworkplaces.ie [accessed 18 March 2011].

Newson, C., Swiderska, C. and Wenban-Smith, J. 2001. *Tourism Without Traffic: A Good Practice Guide*, (London: Transport 2000 Trust).

Newson, C. and Wenban-Smith, J. 1998. *The Healthy Transport Toolkit: A Guide to Reducing Car Trips to NHS Facilities*, (London: Transport 2000 Trust).

Office of National Statistics (Great Britain) and Department for Transport (Great Britain). 2007. *Shopping: Personal Travel Factsheet*, (London: Department for Transport).

Pencheon, D., Cointet, S., Brown, J., Howley, J., Tennision, I., Hannah, G. and Wright, S. 2009. *NHS Carbon Reduction Strategy for England*, (London: NHS Sustainable Development Unit).

Potter, S., Enoch, M. P., Rye, T., Black, C. and Ubbels, B. 2006. The tax treatment of employer commuting support: an international review, *Transport Reviews*, 26(2), 221–237.

Regnerus, H. D., Beunen, R. and Jaarsma, C. F. 2007. Recreational traffic management: the relations between research and implementation, *Transport Policy*, 14, 258–267.

Roby, H. 2009. The death of the travel plan? Paper presented at the *Lunch and Lecture at Loughborough*, Session 6, Loughborough University, Leicestershire, 8 May.

Rye, T. 1998. *Can we Make a Business Case for Employer Transport Plans*? Paper presented to the Car Free Cities Conference, Edinburgh, 22–23 April.

Selby, D, Jones, P., Kagawa, F., Shaw, J., Roberts, L. and Skinner, R. 2009. *Moving Forward: From Travel Plan to Transport Policy?* Occasional Paper 2, (Plymouth: Centre of Sustainable Futures, University of Plymouth).

Sheridan, E. 2008. Tailoring travel plans to the needs of specific audiences: school travel plans. Paper presented at the *Lunch and Lecture at Loughborough*, Session 1, Loughborough University, Leicestershire, 14 March.

Shiftan, Y. and Barlach, Y. 2002. Effect of employment site characteristics on commute mode choice, *Transportation Research Record*, 1781, 19–25.

Steer Davies Gleave. 2001. *Take Up and Effectiveness of Travel Plans and Travel Awareness Campaigns*, (London: Department of the Environment, Local Government and the Regions).

Steer Davies Gleave. 2010. *Station Travel Plan National Pilot Programme: Progress Report*, (London: Association of Train Operating Companies).

Sustainable Development Commission. 2007. *Healthy Future: Sustainable Transport and Active Travel*, (London: Sustainable Development Commission).

Swarbrooke, J. and Horner, S. 2001. *Business Travel and Tourism*, (Oxford: Butterworth Heinemann).

Transport 2000 Trust, Addison and Associates, Matson, L. and Newson, L. 2005. *Making Residential Travel Plans Work: Good Practice Guidelines for New Development*, (London: Department for Transport).

UK Parliament. 2000. *2000 Transport Act*, (London: Crown Copyright). [Online] Available at http://www.legislation.gov.uk/ukpga/2000/38/notes/contents [accessed 5 April 2011].

Van-Ristell, J. A. 2011 Investigating policy impacts on school travel, PhD Thesis, (Loughborough University: School of Civil and Building Engineering).

Wanless, D., Appleby, J., Harrison, A. and Patel, D. 2007. *Our Future Health Secured? A Review of NHS Funding and Performance*, (London: Kings Fund).

Weenen, H. 2000. Towards a vision of a sustainable university, *International Journal of Sustainability in Higher Education*, 1, 20–34.

Chapter 4
Scale: All Travel Plans Great and Small

1 Travel Plans and Scale

As noted from Chapter 3, Travel Plans have now been applied across a full range of segments with varying degrees of success. This chapter will now explore the evolution of the second axis identified at the end of Chapter 2 – scale.

As with the vertical integration discussion outlined previously in Table 2.11, the issue of scale in Travel Planning relates to determining the optimum balance between 'focus' on the one 'small is beautiful' hand, and 'efficiency' on the other 'big is best' hand. Simplistically, in this case at least, the evidence suggests that there are effectively boundaries as to the smallest and largest Travel Plan organisations. Therefore, perhaps a more useful question is to consider how organisations outside of these boundaries might also be included in Travel Planning activities.

This chapter will first examine the influence of organisation size on the performance of Travel Plans in order to establish where these boundaries might be. It will then look at ways of using this information so as to include more organisations within this 'operational range', specifically through the setting up and operation of various types of local Travel Plan Network. Finally, it explores some of the policy decisions that need to be taken at the local authority level relating to how to incorporate yet another 'intermediary' delivery agent.

2 Influence of Organisational Size on Travel Plan Performance

Theoretically it is of course possible for any sized organisation to implement a Travel Plan or at least adopt Travel Plan measures. So, a retail outlet or small manufacturer of one or two people could conceivably provide themselves with bus timetable information and/or apply for a subsidy for a cycle stand or two.[1] However, common sense dictates that the smallest organisations should focus on carrying out their core tasks rather than be sidetracked by the issue of travel because the level of resources required would be too great for the anticipated benefits. At the opposite extreme meanwhile is the grouping of organisations in

1 Interestingly, as already reported in Chapter 3, schemes focused on individual people (known variously as Personalised Journey Planning, Individualised Marketing and so on) have been proven to be effective (albeit sometimes expensive) ways of changing people's travel behaviour. However, such measures are actually delivered at a higher level – usually either at the organisation, network, city or regional level and so are not considered here.

a medium-sized market town of 50,000 or more people, already endowed with a 'conventional' transport planning system. At this level, one would undoubtedly score well on issues of 'scale economies' but there would be a significant question as to whether the benefits afforded by the presence of a 'mediating institution' as a delivery agent of a more typical Travel Plan would continue to apply. In practice, therefore, Travel Plans have generally been implemented by large and medium-sized transport generators.

That said, the evidence is not actually as clear cut as it first appears. In particular, TCRP (2003) highlights a trade-off between targeting large and medium-sized organisations compared with smaller ones. It reports that larger employers are indeed more likely to have a manager in a position whose role could include implementing Travel Plan measures, and can generate a large potential market for those measures once adopted. Conversely, however, smaller employers have fewer layers of decision-making authority to negotiate; are less likely to occupy multiple sites which can be a complicating factor; and face lower overall costs when administering a Travel Plan. In other words, TCRP (2003: 16) finds "no clear consensus on the overall cost-effectiveness of targeting small versus large employers", and summarises the position as follows: "Small employers may be easier to work with, but they only bring in a small number of participants individually. Large employers take longer to implement a program, but they have the potential to bring in many customers".

In recognising these pros and cons, in 2008, Transport for London established a range of programmes to encourage organisations to adopt a Travel Plan, setting key boundaries with different expectations as to the level of involvement expected. Here, companies with 250 or more staff were made eligible for the 'Corporate' scheme, which provided up to £GBP20,000 (€23,000) of professional support and tailored (Travel Plan-related) measures to meet their strategic objectives and business needs in return for a Travel Plan being set up. Meanwhile medium-sized companies could participate in the 'Enterprise' scheme, which involved less subsidy being offered and more generic support in return for the company signing up to a less 'site-specific' but less resource intensive form of Travel Plan (TfL 2010).

Elsewhere, however, it seems that the scale benefits of larger organisations are generally more appreciated. Thus, in Nottingham around 50 of only the largest employers were targeted for membership of the Nottingham Commuter Planners' Club, the idea being that the largest organisations are the 'quick wins', where the most employees can be reached with the least effort by a public authority (Batifois and Fleming 2005). Dutch Vervoer Coordinatie Centrums (VCCs – or Regional Transport Coordination Centres, which have a responsibility for promoting Travel Plans) also hold a remit to target the largest employers first and typically ignore organisations of less than 50 employees (Beljon, R. and Van der Hoef, H. 2003. Personal communication, Transport Officials, Province of South Holland, The Hague, 12 February). And, one further indication of the perceived 'rightness' of the approach has been where regulations have been adopted to impose Travel Plan activity on organisations. So, in Italy only companies larger than 300 employees are required to introduce a Travel Plan, and under Regulation XV in Southern California the lower limit was set at 100 staff members.

Studies considering smaller organisations undertaking Travel Planning are relatively rare, but one early exception is Coleman (2000) which suggests that while in the longer term there may be a role for employers with less than 100 staff adopting Travel Plans, in the short-term the best way forward would be to focus on larger employers. In this she supports even earlier UK work by Rye (1999: 18) which considers firm size and reports:

> Firm size will to an extent dictate the level of resources which can be made available to fund a Travel Plan. It is unlikely that a small-to-medium-sized employer would be able to justify employing on its own a member of staff full-time to work on the implementation and monitoring of a Travel Plan, for example. In comparison, a government transport department or airport will be able to bring considerable resources to bear to ensure the success of its Travel Plan. Measures which require high levels of spending and intervention are therefore more likely to be the preserve of larger employers.

From this, the perceived scale benefits are significant as far as implementing and operating Travel Plans are concerned. Yet it is also true that if only larger organisations are targeted, then a significant proportion of travellers will not be influenced by (Workplace) Travel Plan activities. This is because in the UK in 2006, for example, 52 per cent of the workforce was employed in organisations of more than 50 people. In other words, almost half of the workforce works in organisations where one might assume that Travel Plans might not be deemed appropriate (DBERR and ONS 2007).

Consequently, given the apparent advantages afforded by larger scale Travel Plans, it is unsurprising that various approaches have been tried to broaden their application to smaller organisations through the setting up of local groups or networks. Indeed, this ties in with a further finding of Coleman (2000: 147), which notes: "If smaller businesses are to be targeted they should be done so on an area basis rather than on an individual basis so that resources can be pooled".

In addition, the idea of forming groups or networks to scale up the take up and impact of Travel Plans is supported by Cairns et al (2005), which suggests that more smaller organisations might adopt Travel Planning if it was part of a neighbourhood or area wide approach. And along the same lines, Davison et al (2010) cites an expert Travel Plan consultant who states "Local Travel Plan Networks are always better than conventional Travel Plans, since they involve more employees", once more highlighting a preference for larger potential target markets over smaller and more focused ones. Finally, Rye and MacLeod's 1998 paper suggests that given resource issues facing local authorities then large employers in urban/suburban areas ought to be targeted first, although smaller employers should be tackled in the longer term. Once more, to do this the paper proposes that some form of local group or network that allowed resources to be pooled would be a potential approach given that many small employers feel they are too small to act alone.

The following sections therefore consider how this has been done in practice.

3 Defining Local Travel Plan Networks

In principle, establishing groups or networks of organisations into 'Local Travel Plan Networks' (LTPNs) can provide a mechanism for highlighting the opportunities afforded by Travel Plans and minimising the difficulties of implementing them for organisations, while maximising the transport and wider benefits that such involvement brings to the organisation and the community as a whole. Bluntly, Local Travel Plan Networks can offer:

- Economies of scale – that is networks are collectively able to achieve more than single agencies or employers when dealing with common concerns (thanks to pooled resources delivering higher investment, dedicated staff and greater political influence) and yet allow companies/organisations to focus more on their core competencies.
- Greater flexibility and effectiveness – that is networks have the ability to move Transport Demand Measures (TDM) from a site-specific application to a more flexible and effective area-wide application – given that the nature of transportation and environmental issues is that each employer or agency has the potential to impact upon others – and allow each member to become part of the solution.
- A more strategic and focused approach – that is networks can improve the level of communication between the sectors and allow the level of flexibility necessary to ensure that transport objectives are met in ways that maximise the benefits for businesses, residents and commuters.

Historically, such benefits helped lead to several LTPNs of various forms being set up in North America from the 1980s, before the idea crossed the Atlantic to the Netherlands, then the UK and most recently other parts of Europe, meaning that the majority of the literature reported here derives from these three areas. The next section provides a generic definition of what can be meant by a Local Travel Plan Network.

Energy Efficiency Best Practice Programme (2001: 3) provides a generic definition that: "Travel Plan networks are groups of organisations working together to manage and reduce their car use. Travel Plan networks are run by: local authorities, chambers of commerce, companies, groups of companies and others whose activities generate large volumes of travel".

An alternative definition is provided by Enoch et al (2007b: 5) which states that Local Travel Plan Networks are "a group [or network] of organisations that have come together to share resources and ideas for developing and implementing a Travel Plan in their local area".

4 Categorising Local Travel Plan Networks

Five types of LTPN can be identified from the literature, namely:

- Neighbourhood Transport Forums;
- Area Travel Plans;
- Business Improvement Districts;
- Transport Management Associations; and
- Local Transportation Districts.

These will now be described in turn.

4.1 Neighbourhood Transport Forums

Enoch et al (2007a) describes Area Based Groups (a different term for Neighbourhood Transport Forums (NTFs)) as being informal networks of organisations that operate Travel Plans located within a loosely defined neighbourhood. Such Forums exist where two or more organisations feel combining resources will be a more effective way to deal with transport issues and are generally formed either by local authorities 'suggesting' groups or by one leading organisation taking the lead in helping to address a specific transport issue.

In general, NTFs are multi-member networks and relationships are informal. They also tend to be facilitated by local councils. Full participation is usual and the networks are formed primarily for transport reasons. In the UK, examples are found in Nottingham (the Northside, Southside and Lenton Lane Employer Groups – the latter now 'promoted' to being an Area Travel Plan) (Batifois, H. and Fleming, C. 2005. Personal communication, Transport Officers, Nottingham City Council, Nottingham, 18 November); Bristol (the Temple Quay Employer Group) (Ginger, M. 2005. Personal communication, Transport Officer, Bristol City Council, Bristol, 4 March); and Milton Keynes (TransportMK 2009).

In terms of design, Neighbourhood Transport Forums are the simplest form of LTPN, consisting of informal gatherings of organisations and usually acting primarily as a venue for information exchange which can sometimes lead to measures such as car sharing schemes being adopted (TransportMK 2009). Funding is usually minimal. Local authorities and/or larger organisations within the Forum usually support the events through help with administration, links to other agencies, publicity and venues, while smaller organisations can contribute with publicity or expertise for example.

From an implementation perspective, NTFs are vulnerable for several reasons, namely:

- If transport problems become relatively less important then NTFs become less relevant;
- Their informal nature means they are heavily dependent upon motivated individuals, and should these move on for any reason then the network is potentially in danger as a result;
- They are often not highly resourced and so may be seen as being ineffective at delivering meaningful improvements.

And, given the informality of the arrangements, no formal data on the effectiveness of NTFs in terms of mobility management seems to exist. However, given the low level of resources generally involved it is assumed that impact of these networks is correspondingly low.

For the future, in many ways NTFs help form the first step for organisations with an interest in addressing transport issues (as in the Lenton Lane case in Nottingham). They are thus likely to remain as currently – vehicles for sharing information and coalition building.

4.2 Area Travel Plans

Development zones are local areas developed for a specific use, such as business parks, retail parks, industrial estates, leisure parks and even airports. The overall area is usually owned (or at least managed) by a single private or public sector body that 'hosts' a number of 'tenant' organisations that are located there. Area Travel Plans (ATPs) are local networks that typically form in Development Zone-type environments. The motivations for ATPs being developed are largely similar to those facing the larger individual organisations that set up Travel Plans. With or without local authority support, the site owner or manager (sometimes a tenant) provides a Travel Plan coordinator(s) who establishes, manages and monitors Travel Plans by using contributions, levy or rental fees from the tenants.

Generally, ATP networks consist of multiple members. The relationships are hierarchical and formal, member participation is limited, and the transport element is usually one of several issues dealt with.

Design-wise, ATPs tend to operate as Travel Plans do at large single organisation sites such as large hospitals, universities and large company campuses – the main difference being that several organisations are involved instead of only one. Organisationally, this can make things more complicated, particularly over issues to do with resourcing the Travel Plan. However, there are also cases where this is less of a problem because of the existence of a site management company.

As with large-scale Travel Plans at single organisations, instruments used range from awareness raising 'informational' measures to the provision of alternatives to the car (for example car sharing schemes, improved walking and cycling infrastructure, incentives for using public transport) to measures actively restricting access to the car (for example parking management). For instance, Goudse Port Business Park in the Netherlands established an Area Travel Plan through being involved in European Union project OPTIMUM2. This resulted in a mobility centre, shuttle bus service, mobility card for users, improved cycling facilities and a centralised parking management system being set up (Bossaert and Walvius 2007). Similarly, at the Santa Anna Industrial Area in the County of Bages, Catalonia, measures introduced through the European Union project TRAVEL PLAN PLUS included conducting mobility audits for organisations; promoting travel alternatives to the car (for example through public transport ticketing subsidies); liaising with transport providers in the region; and setting up a joint car and vanpooling programme (TRAVEL PLAN PLUS 2011a).

Significant funding sources tend to include start-up funds from national and/or local government sources, sometimes coupled with developer monies ('captured' through some form of planning agreement) and/or less often contributions from the landlord and tenant. Such support can be in terms of cash, but is also often offered in-kind – perhaps as an employee and office space dedicated to managing the Travel Plan, as help with marketing and publicity, or with Travel Plan expertise. Finance generally of ATPs seems to be offered on rather an ad hoc basis and not as a steady stream. Once again the TRAVEL PLAN PLUS project offers a useful example, whereby a site to the north of Cambridge in the UK used 'kick-start' funding from Intelligent Energy Europe to 'lever-in' funds from several different funding streams within Cambridgeshire County Council and from the private sector. Specifically, money from developers has been used to extend the contracts of two Travel Plan coordinators beyond the timeframe of the original project; a hotel that is due to open in 2012 has offered rent free office space to the network; and several organisations regularly provide in-kind contributions to the network in terms of space and refreshments for meetings and events, and legal advice (TRAVEL PLAN PLUS 2011c).

The key implementation issue for ATPs concerns 'ownership' of the concept by the members within the network. This is because without ownership by the members there is no commitment which in turn means no participation and then no resources. In other words, the temptation for organisations to 'free-ride' – that is to take advantage of the benefits without contributing to the costs – somehow needs to be considered.

In a review of the effectiveness of ten voluntary ATPs in the UK for the Highways Agency, Parsons Brinckerhoff (2008) reports that the average spend by the Highways Agency on each ATP was £GBP60,000 (€69,000), not including monitoring costs (£GBP5,000 to £GBP16,000 which is €5,700 to €18,300 per site) or implementation costs. Interestingly, in terms of performance, no data was presented for two sites. Of the rest, only one site met its target for reducing car use, while at one site the trend was positive, one site missed the target and four sites exhibited negative trends. Results for car sharing, public transport use and cycling were similarly poor, although results for walking were slightly better. No numerical measures of transport behavioural change are provided in the report.

Meanwhile the de Maas office complex in Rotterdam was one of seven office complexes included in a national test project of 'area-oriented approaches' (ATPs) in the Netherlands in the early 1990s (Bakker 2001). Measures introduced included the promotion of carpooling and staff buses; the implementation of strict parking control for individual car commuters; the use of personalised journey planning techniques; and the establishment of a full time transport office. In 1992, the project resulted in car travel to and from the office building falling from 49 per cent to 39 per cent (including carpooling) – amounting to a reduction of 1.3 million car kilometres annually, or a daily reduction of 80–90 cars on the road. Other findings were that even if larger numbers of employees participate in the plan, perhaps through joint systems with other institutions or neighbouring businesses, some 10 to 20 per cent of the costs will still have to be subsidised. In addition, strict parking control is necessary both on- and off-site and the planning, implementation and management of transport works best when supervised by a full time transport manager in the company or office.

Most recently, TRAVEL PLAN PLUS (2011a) reports results from four very different sites across Europe, which in addition to the Industrial Parks in Manresa and the Science and Business Parks in Cambridge already referred to, involved a network of schools from Gyor in Hungary, and 'Cargo City' at Arlanda Airport outside Stockholm in Sweden. This finds that single occupancy car use fell by 6 per cent in Cambridge, 3 per cent in Bages and Gyor and stayed the same in Stockholm, but that the respective figures for vehicle kilometres travelled were 5 per cent up, 9 per cent down, 1 per cent down and 44 per cent up due mostly to significant changes in circumstances at each of the sites.

In general, then, monitoring results from ATPs remain relatively scarce and rather mixed.

Overall, ATPs are still in their infancy but are beginning to mature. Unsurprisingly the critical issue revolves around 'ownership' and especially its link to funding (see earlier). One way of achieving a more stable funding stream is to progress to being a Transportation Management Association (see the section after next). Parsons Brinckerhoff (2008) offers a series of lessons as to the future of ATPs. In particular, it suggests:

- Sites should be selected for reasons that will enable the promoter to meet its strategic objectives.
- At large sites it may prove more effective to divide the site into small, manageable areas, rather than attempting to develop a Travel Plan for the entire site from the outset.
- From the outset, all stakeholders in the development of the ATP should reconcile their motivations and agree a common approach. This can lead to mutually beneficial outcomes.

- Reconciling objectives with realistic targets will improve a Travel Plan's robustness.
- Through working with a sole employer, stakeholder funding can more easily be sourced and funding arrangements can be simpler.
- The focussing of Travel Planning efforts on small areas of larger employment sites to disseminate sustainable travel practices is worth further exploration.
- Each stage of the ATP development process should be documented, including scoping studies and the reasons why the site was selected.
- The approach to recording baseline and future monitoring data should be consistent to ensure that data sets are comparable.
- Travel Plans are not only about promoting and incentivising the use of sustainable modes; they can include disincentives such as parking management.
- Working closely with public transport operators may enable arranging for discounted tickets to be offered to staff of businesses in the local area.
- A reasonable degree of mode shift can be achieved through promoting existing facilities and benefits of alternative modes.
- Voluntary Travel Plans should aim to maximise the potential of existing provision for sustainable transport.
- A car-sharing scheme can be successful in reducing SOV use if it is well promoted.
- A Continuation Strategy should be developed early on the process of the development of the ATP, with a plan for the handover of leadership.

4.3 Business Improvement Districts

"Business Improvement Districts (BIDs) are self-assessment districts formed by property or business owners" (Morcol and Zimmermann 2006a: 6). Most BIDs are self-taxing districts that are enabled by state legislation and legally established by municipal governments.

BIDs operate in a range of different legal frameworks across the USA and beyond and are also known by various other names including downtown improvement districts, neighbourhood improvement districts, community improvement districts, community benefit districts, self-help business improvement districts, self-supported municipal improvement districts, business improvement areas and city improvement districts. Shoup (2004) reports that the first BID was established in downtown Toronto in 1965. Hoyt (2004) adds that BIDs began operating in Canada, New Zealand, the United States and South Africa and have more recently spread to other regions of the globe including Australia, Austria, Belgium, Denmark, France, Germany, Japan, Netherlands, Norway, Portugal, Spain, Sweden, and the United Kingdom, where they began operating from the beginning of 2005 and now number more than a hundred (UK BIDs 2011).

BIDs are typically multi-member organisations that have non-hierarchical formal relationships. BIDs enjoy significant participation from their members and are formed for many reasons of which transport is sometimes one.

Operationally, Wolf (2006) reports that of four BIDs studied in Washington DC, one (in Georgetown) has transport play a significant role while another (the Downtown BID) provides minor transport services, and the others ignore transport altogether. More generally, Morcol and Zimmermann (2006a) reports the main functions of a BID as being of four types: business services, policy advocacy, 'traditional' public services and 'comprehensive governmental authority'. The main source of revenues for most BIDs is assessments levied on businesses which are typically based on property values (Morcol and Zimmermann 2006a).

In some ways the implementation challenges facing BIDs are focused on establishing the BID in the first place – usually as a result of winning a ballot of potential member organisations – at least for the set period of time until the next ballot is called (usually after five years). This is due to the guaranteed source of revenue thanks to the legal requirement that BID member organisations contribute through a surcharge on business rates. In this case then, from a transport planner's perspective the task is to ensure that transport improvements are seen by the rest of the BID team to be 'worthy' of supporting. This is probably easiest to do before the ballot occurs although there are instances (for instance with the Better Bankside BID in central London – see Better Bankside 2009), where transport became a priority after the ballot had taken place (Tyler et al 2007).

Looking in more detail at the Better Bankside Travel Planning Group (BBTPG), this was set up in November 2005 in central London, south of the River Thames from the financial district (TRAVEL PLAN PLUS 2011b). Overall, the BID covers an area of two by one-and-a-half kilometres and includes a range of organisations within its 460 members, each of which pay an annual levy. One of seven 'theme groups' made up of roughly 15 regular members, the BBTPG also includes representatives from the London Borough of Southwark and Transport for London, and oversees services including:

- free cycle maintenance sessions and maintenance training;
- a regular employee travel survey;
- the development of secure cycle parking facilities;
- a walking programme;
- public transport information and lobbying activities, and
- a cycle freight pilot project.

All of these activities take place under the umbrella of a 'Master Travel Plan'. The first version developed in 2006 included setting targets for increasing the proportion of commuting and business trips within London on foot and by bicycle by 2010. Data collected as part of an Employee Travel Survey in 2010–11, have shown that these targets have been met and new targets will be set for the period 2011 to 2015.

The main advantages of BIDs relate to their independence from local government meaning that political risk to politicians "is reduced because residents are not taxed to pay for them, meaning any 'public' money that subsequently emerges seems 'free'" (Morcol and Zimmermann 2006b: 89). On the downside, there is a lack of public accountability to their local governments, and to property owners in their districts. Crucially, BIDs can be very successful at leveraging funds from governmental sources (Morcol and Zimmermann 2006a).

Once again, evidence of effectiveness is scarce, particularly regarding transport impacts, although evidence from broadly similar Community Improvement Districts in Atlanta, Georgia suggests stakeholders perceive them as being very effective (Morcol and Zimmermann 2006b).

For BIDs in the future, the findings of Morcol and Zimmermann (2006a: 13) are that "BIDs are becoming less like special-purpose entities and more like general purpose governments, as they take on larger and more governmental powers and deliver a broader range of services. They are also gradually gaining more legal and de facto autonomy from the local governments that created them". Clearly then, such a shift potentially has significant ramifications, particularly in terms of the reduced accountability of the private sector to the public, and so care must be taken to ensure that the needs of those without a direct involvement in such an organisation are not ignored.

4.4 Transportation Management Associations/Organisations

Ferguson (2007) states that alternative names for TMAs include Transportation Management Organisations, Transportation Management Initiatives and Transportation Management Districts (TMDs). Specifically, he notes that TMOs tend to be more action oriented than TMAs, TMIs refer to TMA start ups and TMDs refer to special purpose tax districts and/or operating areas. Schreffler (1986) notes that TMOs are generally private, not for profit corporations formed so that employers, developers and/or retailers can collectively address transport-related problems. He continues that several characteristics of TMOs can be cited.

- Initiated by the private sector;
- Possess non-profit corporate status;
- Place decision-making in the hands of member representatives;
- Maintain small staffs of planners and liaison personnel;
- Are funded by membership fees and assessments.

Perhaps more broadly, Enoch et al (2007b) notes that TMAs tend to be multi-member, non-hierarchical formal relationships, with full/limited member participation and are formed primarily for transport reasons.

Taking a design perspective, Hendricks (2004: 130) draws on a North American-wide study to quantify this further, and reports that:

> TMAs are either regional, suburban or corridor in geographic scope and serve an average of 48,500 commuters. The central focus of the TMA would be policy leadership, advocacy, and service provision. It would have 20 to 80 members composed mainly of business employers. The TMA would use contact from the executive director and peer to peer contact as its chief member recruiting tactic.

Ferguson (1997) notes that TMAs provide a wide variety of products and services, the most common being transit information and ride-sharing assistance. ACT (2010), Miller et al (2005) and Ferguson (2007) expand these, and note that TMAs may serve as:

- Promoter – advocacy, marketing and promotions.
- Broker – brokerage and referral services (for example for parking and shuttle buses – so as to make more efficient use of resources).
- Provider – direct provider of corporate (such as consultancy) and/or commuter services, focusing on the exchange of information and other more tangible incentives.
- Owner/operator – facility and equipment manager, such as parking, vanpools, shuttle.
- Regulator.
- Educator.
- Facilitator – for example by creating a forum for consensus building.

On finance, Hendricks (2004: 131) states that:

> in 1993, 20 per cent of respondents obtained 100 per cent of their funding from dues. Dues made up 47 per cent of average total revenue. In 2003, 5 per cent of respondents obtained 100 per cent of funding from dues, and [overall] dues made up 40 per cent of average total revenue.

Other funding sources include transit fares, private grants, taxes, municipal sponsors, the parent organisation, foundation grants, vanpool revenues, promotional events, parking fees and company investments. Miller et al (2005) states that most TMAs now require subsidies, especially for start-up costs.

By contrast, in Canada, only two of the eight TMAs use member dues (and then as a minor source of income) while the majority of funding is provided through government grants (Hendricks 2004). Similarly in the UK only one TMA (Dyce TMO in Aberdeen) has so far been able to introduce membership fees though several others have attempted to do so (Contemporary Transport 2009) while the remainder still rely on grants.

Ferguson (1997) states four steps to TMA implementation. First, is TMA formation. Second, is the development of a mission for the TMA including long-term goals and short-term objectives. Third, is to develop a work programme, including products and services to be offered and fourth is to develop an evaluation plan. In terms of barriers to the process, Litman (2010: 2) finds the main barriers to setting up and operating a TMA "are a lack of support among stakeholders and often the perception that short-term benefits are small if there is no immediate parking or traffic congestion problem". Meanwhile Diggins and Schreffler (1992) cites financial assistance and one-to-one information assistance as being the primary assistance needs for new TMAs. The 'free-rider' problem, whereby organisations do not join the TMA but benefit from its services nonetheless, is also often seen as being problematic.

Yet again, the main point on performance is the scarcity of evidence. Thus, in the ACT (2010) *Handbook of TMAs* the only reference to transport impacts suggests that single occupancy car use typically fell by between 2 and 5 per cent. And a review of 58 transportation control measures where TMAs were found to be "relatively effective and cost effective when compared with many of the other projects evaluated" (Pansing et al 1998: 101), only two TMA schemes were evaluated which "potentially skewed the results".

Two exceptions from sites where data is available are as follows. First, in 2001 member companies in the Amsterdam Schiphol Airport TMA, a partnership between the Dutch central government, airport operator, airport-based companies and public transport operators, employed 42,300 employees, or 80 percent of the total workforce of Schiphol-based companies. Total car use, including car sharing, reduced from 72 percent in 1996 to just under 70 percent in 2000/1 and total public transport use increased from 19 percent to 21 percent over the same period. Schiphol airport regarded this as a success, as car use in society as a whole has risen during this period (Reeven et al 2003; Sam 2001). Second, in a survey of employers in Atlanta, Zuehlke and Guensler (2007) reports TMA members (15 per cent of employers) as achieving higher levels of implementation and having more positive employer perception of employer trip reduction programmes than non-members.

As with BIDs, Mallett (1993: 14) foresees that in future "conceivably TMAs could develop from specific service agents to more general providers by being involved in land use decisions, providing child care facilities, and as a forum for the discussion of public policy issues". Ferguson (2007), however, concludes that TMAs are unlikely to grow as an alternative form of governmental entity but should retain an important niche as a provider of TDM service providers. Meanwhile, Hendricks (2004) reports that a general decrease in involvement from TMA members (member dues, fewer board meetings, less volunteerism) and increase in work by professional staff together with increase in government funding could be seen as being negative developments. This is because government as a TMA customer represents a less specified, more nebulous target market than developers and business owners.

In looking at what TMAs can offer Europe, Benard and Cre (2007) suggests that there is a need for coordinating structures that can deliver a range of mobility management measures, though less than are necessary in the US context.

Specifically, it points out that infrastructural, operational public transport and planning activities would not be required in [continental] European TMAs.

Overall, TMAs provide a relatively formal means of delivering a LTPN, but there is one category that takes this formality to the next level and indeed effectively returns the responsibility from the organisational realm to the public, albeit in a modified way. This is the Local Transportation District.

4.5 Local Transportation Districts

There are several types of Local Transportation Districts covered within this category. These include Parking Benefit or Management Districts, whereby a proportion of the revenue from parking charges and fines is earmarked to be spent on transportation improvements in the district where the monies are collected (Shoup 2004), the Transportation Improvement District (Mallett 1993) and Transportation Management Districts (TMD's) (Pogue 1997).

The key point is that legal ordinances are used within a legally designated geographical area either to set up a form of tax or charge (for example on parking spaces) in order to raise revenues to pay for improved transport, or else to require organisations to perhaps produce, implement and monitor some form of Travel Plan. In other words, Transportation Management Districts form a mechanism for enacting TDM regulations, thus providing the motivation for businesses to participate (Flynn and Glazer 1989; Pogue 1997). For example, in Montgomery County, Maryland, Flynn and Glazer (1989) notes that the TMD requires all existing and new employers to file TDM plans that seek to meet an average auto occupancy of 1.3 persons per vehicle and 25 to 30 per cent public transport ridership. For further instances of institutional arrangements whereby local charges or taxes (for example from employment taxes, road or bridge tolls, sales taxes and so on) see the book *Unfare Solutions* (Ubbels et al 2004).

LTDs are administratively probably the most complex type of LTPN to set up and the most resource intensive. This is because they are effectively a government body, unlike the other network types described here. However, they can also potentially be the most effective in delivering transport goals due to the greater range of instruments available.

Transportation Management Districts are probably the closest related existing LTDs to Travel Plans, and the only TMDs in operation currently exist in Montgomery County, Maryland. Here, approximately 120,000 commuters and 1,120 employers are arranged in four TMDs (soon to increase to five), that range in size from 5,000 to 65,000 employees – 50 to 520 employers. TMDs legally require employers of more than a set minimum of employees to produce, implement and monitor a Travel Plan (MCC 2004). These TMDs offer employers a truly comprehensive package of measures. On the one hand, employers are legally required to adopt Travel Plan measures, but on the other the TMDs provide a huge range of free services to help employers market alternative transport options to employees (MCCS 2009), focusing mainly on the provision of information.

One significant attraction of LTDs is that funding sources are generally established from the start and are often relatively secure. Thus, Special Parking Districts are often funded from parking charges and parking fine revenues (Shoup 2004). For the Montgomery County TMDs (Pogue 1997), dedicated funding sources considered included various parking taxes and charges, membership fees, developer levies and other new taxes such as an employee head tax or payroll tax.

Implementing some form of Travel Plan focused LTD (at least outside of the United States where legal frameworks are broadly sympathetic to such entities) is likely to be complicated. This is because, in general, such bodies – equipped with specialised legal powers and dedicated sources of funding – have not been implemented on any scale meaning that significant political, legal and institutional barriers would need to be overcome. This would seem to be an unlikely prospect unless it could be demonstrated that the Montgomery County examples were delivering significantly more impressive results than the alternative network types.

Performance-wise, Pogue (1997) reports the North Bethesda TMD as being an effective tool for reducing energy consumption, lowering emissions and changing travel behaviour. Crucially, this depended on the support of elected officials and the community at large, and secondly it must co-opt private employers to cooperate with it to produce measurable improvements. While no results could be determined for the TMDs, there are a series of commuting goals that have been identified for each TMD (MCCS 2009). These goals represent the percentage of commuters not driving to work during the most congested times of the day and for the United States are stringent, ranging from 37 per cent to 50 per cent. More generally Mallett (1993) asserts that special district governments [including Local Transportation Districts] do help provide improved services in areas with limited financial and administrative capabilities and are generally independent of local government (namely shielded from politics and political abuses). However, they can also be criticised for being inefficient, unfair, confusing to citizens and unaccountable.

As noted above, implementing a LTD would seem to be an unlikely prospect without some significant shift in the current context. Nevertheless, the high level of ambition signalled in Montgomery County does mark out the TMD as a target to potentially strive towards.

5 Selecting a Suitable Type of LTPN

Table 4.1 summarises the characteristics of the LTPN types discussed.

Table 4.1 Local Travel Plan Network type attributes

	Neighbourhood Travel Forums (NTFs)	Area Travel Plans (ATPs)	Business Improvement Districts (BIDs)	Transportation Management Associations (TMAs)	Local Transportation Districts (LTDs)
Definition	Informal networks in a loosely defined neighbourhood.	Local areas developed for specific uses.	LA-business partnership to invest within a defined area.	Private, non-profit, member-controlled organisations for defined area.	Companies in defined area legally required to develop Travel Plans.
Network structure	Organisations all equal.	Leading organisation and members.	Coordinating organisation created.	Coordinating organisation created.	Led by LA coordinating organisation.
Leader-member relationship	Common interest – informal.	Landlord-tenant – formal.	Financial (tax) – very formal.	Financial (member fee) – fairly formal.	Legal requirement – very formal.
Power structure	Power shared equally.	Landlord in control.	Membership in control.	Membership in control.	LA in control.
Transport only issue?	Yes	Yes	No	Yes	Yes
Primary actor	LA/private companies.	DZ/LA	LA initially, then private BID company.	LA/private companies.	LA
Secondary actors	Private companies/LA.	LA/DZ	Private company members.	Private company members.	Private companies.
Role of local authority (voluntary Travel Plan)	Support	Support	Initiator and facilitator.	Support	N/A
Role of local authority (mandatory Travel Plan)	Support	Regulator	N/A	Regulator	Regulator
Funding	Ad hoc grants, scheme basis.	Ad hoc grants, rent.	Business levy.	Ad hoc.	Local authority funded.

Source: Adapted from Enoch et al (2007b)

The next step for organisations in selecting a type of LTPN suitable for a particular site then would be for them to take account of the contextual circumstances. Enoch at al (2007a) suggests these include the primary actors, the motivations for group/network formation, the perceived scope and scale of the problem and the organisational environments, along with geographic, political, and institutional factors.

5.1 Primary Actors

Traditionally transport policy is implemented by suppliers (operators, construction companies, technology providers, infrastructure managers and vehicle producers) and by various levels and agencies of government, while the field of Travel Planning involves a number of new actors. These include generators of travel such as workplaces, visitor attractions and schools, as well as private developers. With the establishment of a LTPN it is often the case that a new actor is either formed (as in the case of a TMA) or becomes included in delivering transport (for example an existing business network or site management company). It is also interesting to examine which actor tends to take the lead where each type of LTPN is concerned.

Thus, with ATPs in the UK local authorities and bodies such as Transport for London and the national trunk road operator the Highways Agency have tended to try and persuade landowners, managers, tenants and developers of business/retail zones to participate. This is usually done through specially contracted expert consultants, who then may be invited to operate the ATP whilst it becomes established.

Neighbourhood Transport Forums meanwhile are often facilitated and sometimes led by local authorities through the provision of basic funding and administrative resources together with links to public transport operators and public service and information providers.

Diggins and Schreffler (1992), Mallett (1993) and Ferguson (1997) note that TMAs can be organised by two main groups – either by developers (as a means of securing planning permission from the local authority or as a marketing tool for potential tenants) or in the majority of cases by major landowners or major businesses. Morcal and Zimmermann (2005) notes some TMAs are created voluntarily by private sector organisations but that others are encouraged, facilitated or even mandated by local authorities, while Miller et al (2005) asserts that TMAs are formed by regional or local governments, chambers of commerce, clustered businesses or management of a major employer such as a hospital or university.

5.2 Motivations for Group Formation

As reported in Chapter 2, DeHart Davis and Guensler (2005) reports that employers voluntarily mediate policy problems either because of:

- Organisational self-interest (that is when there is perceived to be a direct or indirect economic payoff);
- Organisational control (that is where actions taken reduce employee discretion so that employees behaviour becomes more closely aligned to the goals of the organisation);
- Association membership (where peer pressure generates a socialisation process whereby identity is developed, norms are communicated and members then conform to these norms so as to gain the approval and respect of fellow members).

Clearly, LTPNs tap most obviously into the final motivation although there may also be links with the first.

Looking at individual types of LTPN, perhaps the crucial relationship concerns not the type of motivation, but the degree. So, weakly motivated organisations will be more likely to join a NTF or ATP, while networks seeking more tangible benefits might gravitate towards the BIDs, TMAs and LTDs.

5.3 Perceived Scope and Scale of the Problem

In simple terms, transport-focused networks (ATPs, NTFs, TMAs, LTDs) will more likely be established in locations with significant transport issues whereas where other issues are dominant it may be that more generalised networks such as BIDs are formed (which may or may not then seek to address transport problems). Next, the scale of the problem refers to whether the issue to be dealt with requires a localised or site-based solution, or whether it needs to be addressed on a neighbourhood basis. ATPs and TMAs tend to focus on site-specific concerns, whereas NTFs, BIDs and LTDs are less delineated and deal with wider neighbourhood-level issues.

5.4 Organisational Environment

Obviously the organisational environment – which refers to the number of organisations, locations (edge of town, city centre), distribution (clusters, evenly spread, corridors), sectors (industrial, leisure, retail, commercial, health, education, mixed) and size (number of employees and visitors) – is another crucial element in determining the type of LTPN adopted. For instance, as with the perceived scope and scale of the problem ATPs and TMAs might be appropriate in a business or retail park context whereas NTFs, BIDs and LTDs are usually less restricted by site boundaries. From a different perspective, organisations with fewer resources and/or less motivation to address transport issues are more likely to develop NTFs and ATPs than TMAs, BIDs or LTDs.

5.4 Geographic, Political and Institutional Factors

Regarding geographic factors, the ATPs and TMAs tend to have distinctive borders and are typically located at edge or out-of-town sites, whereas ABDs, BID's and the TMD examples are usually found in inner city or downtown areas and can have quite blurred boundaries. However, these generalisations do not always hold true. Litman (2010: 3) suggests "TMAs are appropriate for any geographic area where there are multiple employers or business that are clustered together, which can benefit from cooperative transportation management or parking brokerage services. Regional and local governments, business associations and individual businesses can all help establish TMAs".

Politically, the split has less to do with the type of network than with the motivation behind its formation – that is: is it implemented because of self-interest or as a legal requirement? The exceptions are that LTDs are always pushed by legal requirement while the NTFs are usually voluntary. Of the remainder, ATPs can be either voluntary or mandatory, while the BIDs and TMAs seem far more dependent on both public and private bodies 'buying-in' to the groups and taking the lead at different stages of their development. Legally, NTFs and ATPs can exist anywhere, as can TMAs although some advice may be necessary as to the most effective administrative form. BIDs and LTDs require legislation before they can be introduced if none is currently in place. For example, TMDs are not legally sanctioned anywhere outside of Montgomery County, Maryland in the USA.

The institutional issue is focused on whether a suitable existing group may be used to 'piggyback' transport issues. This can be easier than setting up a brand new specialist group but can also be less focused on delivering transport goals, especially if transport objectives are not fully accepted by other members. BIDs form the pre-existing LTPN types, while ATPs, NTFs, TMAs and LTDs are set up specifically to deal with transport issues.

6 Practical Lessons for Developing LTPNs

There are several studies that present practical guidelines for implementing LTPNs. First, findings from the Niches Project (Niches undated) break down the implementation phase into three parts – preparation, implementation and operation – and state that practitioners should:

Preparation

- Devise a good strategic plan and a winning idea;
- Have a common understanding of the problems and appropriate solutions within the selected area;
- Create core membership, steering committee and local champion;
- Have a well-specified geographic scope and travel market;
- Ensure good cooperation and interest from a wide range of stakeholders;
- Establish commitment for broad and sustained funding (public and private);
- Develop good complementarities between existing services and new services to be provided;

Implementation

- Set up a well-defined structure;
- Ensure marketing and membership development are considered;
- Aim for early success experience;
- Make tailor-made services according to the needs of the travel market;

- Ensure good vertical, horizontal and spatial integration and adequate development orientation;
- Generate interest and awareness among target groups;
- Provide board officer training;

Operation

- Ensure effective public private partnerships and stakeholder involvement;
- Develop sustained funding streams; and
- Measure and report results.

Second, Enoch et al (2007b) sets out how to assess the suitability of what type of LTPN to apply in a particular area and then explains how to establish a network in terms of identifying a lead business organisation, making a business case, creating the partnership, consulting members and so on. Similarly, ACT (2010) provides details on studying the feasibility of forming LTPNs and on enhancing success once a network is operational.

Finally, TRAVEL PLAN PLUS (2011b) sets out numerous recommendations for each of six steps that run from analysing the needs and feasibility, to selecting the LTPN type, preparation, implementation, monitoring/reviewing and finally evaluation.

7 Accommodating LTPNs from a Local Authority Perspective

In accommodating LTPNs within the policy framework, Contemporary Transport (2009) highlights five quite specific actions for policy makers that would help LTPNs become more widespread and effective. These are:

- Create a road show workshop outreach campaign.
- Develop a recognised guidance manual.
- Establish a funding scheme for LTPNs.
- Consider how to promote transport measures in pre-existing networks with wider operational remits (as with BIDs for example).
- Evaluate existing LTPN performance.

Taking these ideas forward somewhat, TRAVEL PLAN PLUS (2011c) then suggests that three levels need to be considered, namely local/regional, national and European. Thus, at the local level:

- A LTPN should only be pursued where the organisational and site contexts mean they are an appropriate measure.
- LTPNs should ideally form part of the planning process for new developments.

- LTPNs may require strong leadership initially from policy makers, but that over time it is strongly advised that LTPN members be increasingly involved and empowered once the network is up and running.
- Policy makers should try and consider the needs and objectives of LTPNs when designing related transport and wider policy measures.

Nationally, the need is for:

- LTPN champions to be charged with 'spreading the word'.
- The continual assessment of LTPN benefits on an actor by actor basis.
- Regular reinforcement of the LTPN concept through meetings with relevant actors, policy guidance notes and embedding in government policy frameworks as far as possible.

Finally European (supra-national) policy makers ought to:

- Promote the adoption of LTPNs as good practice in terms of transport and wider policy goals.
- Establish funding streams to help kickstart new networks.
- Provide additional research opportunities for assessing LTPN benefits.
- Facilitate dissemination opportunities for LTPNs.

8 The Future of LTPNs

For the future, EEBPP (2001) notes that the benefits of expanding networks can be significant because:

- larger networks offer economies of scale and the ability to spread costs more widely;
- representing larger numbers of employers (and employees) strengthens a network's position in negotiations with policy and decision-making bodies, such as local authorities and public transport operators;
- the larger the network, the greater the pool of 'good practice' information that members have to share;
- the larger the membership, the more viable some services become, such as car-sharing schemes;
- the scope for special-interest sub-groups increases with network size;
- within the network catchment area, the greater the proportion of organisations that are members of that network, the less the influence of non-members who may be 'dissenters'.

It also suggests that funding be sought from a range of sources so as to minimise dependence on a single provider.

Meanwhile, in looking at the potential of LTPNs in a review of the opinions of Travel Plan experts from across Europe, Davison et al (2010) reports that LTPNs are deemed to represent an effective and cost effective way of extending the benefits of Travel Plans, and can be expected to play an increasing role in future. It also points out that while there are additional challenges to traditional Travel Plans due primarily to the bringing together of multiple organisational cultures and objectives, these could be addressed through effective communication. Moreover, it suggests that the advantages of increased financial and lobbying power are seen as providing sufficient additional benefit to justify the extra effort involved. The paper also reports that LTPNs are suited to central business districts and business parks where there are many organisations, where there is a pre-existing transport infrastructure supporting choice, and where space is at a premium and organisations are likely to experience shared problems (and so would be looking for shared solutions). In addition, good pre-existing relationships can help, while the most effective LTPNs also tend to be problem driven. Finally, regarding the optimal size of a LTPN, Davison et al (2010) notes that the balance needed is to be large enough to deliver economies of scale but small enough as to be manageable.

9 Conclusions

In conclusion, in general the larger the Travel Plan organisation the greater the potential for Travel Plans to be implemented and to achieve their objectives. However, as reported earlier there are some issues with increasing scale, and consequently one mechanism for trying to achieve the benefits of both a larger scale and smaller scale Travel Plan, is to establish a Local Travel Plan Network of some form.

Such networks range from informal, loosely defined networks drawing on minimal resources to more formal and legally defined organisations with access to a series of possible promotional, regulatory and fiscal tools. They can be either transport-focused or more generally aligned and are heavily influenced by contextual factors such as the primary actors, the motivations for group formation, the perceived scope and scale of the problem and the organisational environments, along with geographic, political and institutional factors.

In terms of performance, there appears to be a direct correlation between the type of LTPN (and the corresponding level of resources available) and the contribution the network can make to reducing transport externalities. There is also an underlying story that LTPNs have not yet realised their full potential either at the individual local network level or at the wider community transport system level. As a result, LTPNs like Travel Plans more generally, are far more likely to be implemented (and to operate effectively) if supported by stakeholders additional to the immediate 'delivery agents' – for example government, public transport operators and the wider Travel Plan industry – who share similar and/or complementary goals. Such support can be offered in a variety of ways and it is these that are explored in Chapter 6.

However, before such support measures are examined, it is necessary to look again at the product that needs to be supported – in other words, at how the Travel Plan functions within the structure of organisations (and indeed of LTPNs).

BAA, the operator of Stansted Airport has established an Area Travel Plan to help its member organisations get staff, suppliers and customers to the site more efficiently and effectively

LTPNs must involve a high level of networking if they are to function properly

References

Association for Commuter Transportation. 2010. *TMA Handbook: A Guide to Successful Transportation Management Associations*, (Washington DC: Association for Commuter Transportation).

Bakker, M. 2001. Rotterdam: transport demand management in the De Maas Office Complex, European Academy of the Urban Environment, 23 August. [Online] Available at http://www.eaue.de [accessed 26 May 2009].

Benard, V. and Cre, I. 2007. *Transportation management associations: what is in it for Europe?* Paper presented at the Polis Annual Conference, Toulouse, France, 15–16 March.

Better Bankside. 2009. Better bankside master travel plan. [Online] Available at http://www.betterbankside.co.uk/travelplan [Accessed 6 May 2009].

Bossaert, E. and Walvius, M. 2007. Improvement accessibility of business park Goudse Port in the City of Gouda, the Netherlands, ELTIS – European Local Transport Information Service, 15 October. [Online] Available http://www.eltis.org [accessed 27 March 2009].

Cairns, S., Sloman, L., Newson, C., Anable, J., Kirkbride, A. and Goodwin, P. 2005. *Making Smarter Choices Work*, (London: Department for Transport).

Contemporary Transport. 2009. *TMA Study*, Unpublished Report, (London: Transport for London).

Davison, L. J., Enoch, M. P. and Ison, S. G. 2010. *European experience of travel plans: an expert perspective*. Paper presented to the World Conference on Transport Research, Paper 1286, Session G4.3, 12–15 July, Lisbon.

DeHart Davis, L. and Guensler, R. 2005. Employers as mediating institutions for public policy: the case of commute options programs, *The Policy Studies Journal*, 33(4), 675–697.

Department of Business, Enterprise and Regulatory Reform (Great Britain) and Office of National Statistics (Great Britain). 2007. *Small and Medium-sized Enterprise (SME) Statistics for the UK 2006*, Statistical Press Release URN 07/92, (London: Department of Business, Enterprise and Regulatory Reform).

Diggins, L. and Schreffler, E. N. 1992. Status report on transportation demand management in California, *Transportation Research Record*, 1346, 53–61.

Energy Efficiency Best Practice Programme. 2001. *A Guide on How to Set Up and Run Travel Plan Networks*, Good practice guide 314, Energy Efficiency Best Practice Programme, (London: The Stationery Office).

Enoch, M. P., Zhang, L. and Ison, S. G. 2007a. Unlocking the potential of site based mobility management through local travel plan groups, *World Transport Policy and Practice*, 13(2), 23–39.

Enoch, M. P., Tyler, S., Zhang, L. and Pearce, H. 2007b. *Local Travel Plan Groups: A Practical Guide to Setting up an Effective Travel Plan*, (London: SEA/Renue on behalf of Transport for London).

Ferguson, E. 1997. Privatisation as choice probability, policy process and programme outcome: the case of transportation management associations, *Transportation Research Part A*, 31(5), 353–364.

Ferguson, E. 2007. Transportation management associations: a reappraisal, *Journal of Public Transportation*, 10(4), 1–26.

Flynn, C. P. and Glazer, L. J. 1989. Ten cities' strategies for transportation demand management, *Transportation Research Record*, 1212, 11–23.

Hendricks, S. 2004. Results of 2003 transportation management association survey, *Transportation Research Record*, 1864, 129–134.

Hoyt, L. 2004. *The Business Improvement District: An Internationally Diffused Approach for Revitalization*, (Washington DC: International Downtown Association).

Litman, T. 2010. Transportation Management Associations, Online Transportation Demand Management Encyclopaedia, Victoria Transport Policy Institute, 26 January. [Online] Available at http://www.vtpi.org [accessed 4 April 2011].

Mallett, W. J. 1993. Private government formation in the DC metropolitan area, *Growth and Change*, 24(3), 385–416.

Miller, J. H., Hough, J. and Hellegaard, L. 2005. Asset management and funding: transport and non-profit institutions, in the *Handbook of Transport Strategy, Policy and Institutions*, edited by K. J. Button and D. Hensher. Handbooks in Transport: Volume 6, (Amsterdam: Elsevier), 391–402.

Montgomery County Council. 2004. *Friendship Heights Transportation Management District Montgomery County, Maryland Activities and Performance Fiscal Years 2000–2002*, (Silver Spring: Montgomery County Commuter Services).

Montgomery County Commuter Services. 2009. Montgomery County Commuter Services homepage. [Online] Available at http://www.montgomerycountymd.gov/ [accessed 6 May 2009].

Morcol, G. and Zimmermann, U. 2005. Transportation management associations: prospects and problems for public administration, *International Journal of Public Administration*, 28, 1095–1116.

Morcol, G. and Zimmermann, U. 2006a. Metropolitan governance and business improvement districts, *International Journal of Public Administration*, 29, 5–29.

Morcol, G. and Zimmermann, U. 2006b. Community improvement districts in metropolitan Atlanta, *International Journal of Public Administration*, 29, 77–105.

National Cooperative Highway Research Program. 2010. *State Department of Transportation Role in the Implementation of Transportation Demand Management Programs*, Research Results Digest 348, (Washington DC: Transportation Research Board).

Niches (undated) Transportation Management Associations (TMAs), Innovative demand management strategies Policy Notes, Niches, EC Sixth Framework

Coordination Action, Brussels. [Online] Available at http://www.niches-transport.org [accessed 4 April 2011].

Pansing, C., Schreffler, E. N. and Sillings, M. A. 1998. Comparative evaluation of the cost-effectiveness of 58 transportation control measures, *Transportation Research Record*, 1100, 97–104.

Parsons Brinckerhoff. 2008. *Effectiveness of the Highways Agency's Area Travel Plans*, Final Report, (London: Highways Agency).

Pogue, T. D. 1997. The transportation management district, *Energy Policy*, 25, 1227–1234.

Reeven, P. V., Vlieger, J. D. and Karamychev, V. 2003. *BOB Airport Accessibility Pilot*, Final Report, (Rotterdam: Erasmus University Rotterdam Transport Economics).

Rye, T. 1999. Employer transport plans – a case for regulation, *Transport Reviews*, 19(1), 13–31.

Rye, T. and MacLeod, M. 1998. An investigation of employer attitudes to employer transport plans. Paper presented at the 26th *European Transport Conference*, Loughborough, UK, 14–16 September.

Sam, F. 2001. Public-private cooperation gives benefits for companies and employees, VCC Schiphol, Airport Amsterdam Schiphol, Paper presented at an *EPOMM Workshop,* Milan, 14 December.

Shoup, D. 2004. *The High Cost of Free Parking*, (Chicago: American Planning Association).

Schreffler, E. N. 1986. Transportation management organisations: an emerging public/private partnership, *Transportation Planning and Technology*, 10(4), 257–266.

TRAVEL PLAN PLUS 2011a. *Local Travel Plan Networks: Final Implementation and Evaluation Report*, Work Packages 4 and 5, TRAVEL PLAN PLUS Project IEE-07-592, (Brussels: EC-EACI-IEE-STEER Programme). [Online] Available online at http://www.travelplanplus.eu [accessed 2 May 2011].

TRAVEL PLAN PLUS. 2011b. *Local Travel Plan Networks: Recommendations for Policy Makers*, Work Package 6, TRAVEL PLAN PLUS Project IEE-07-592, (Brussels: EC-EACI-IEE-STEER Programme). [Online] Available online at http://www.travelplanplus.eu [accessed 2 May 2011].

TRAVEL PLAN PLUS. 2011c. *Local Travel Plan Networks: A Practical Guide to Implementation*, Work Package 6, TRAVEL PLAN PLUS Project IEE-07-592, (Brussels: EC-EACI-IEE-STEER Programme). [Online] Available online at http://www.travelplanplus.eu [accessed 2 May 2011].

Transit Cooperative Research Program. 2003. *Strategies for Increasing the Effectiveness of Commuter Benefits Programs*, TCRP Report 87, (Washington DC: Transportation Research Board).

Transport for London. 2010. *A New Way to Work*, Transport for London. [Online] Available at http://www.tfl.gov.uk [accessed 18 August 2010].

TransportMK. 2009. Transport Partnership MK homepage. [Online] Available at http://www.mkweb.co.uk/transportmk/home.asp [accessed 6 May 2009].

Tyler, S., Enoch, M. P. and Zhang, L., 2007. *Setting up local travel plan groups: the future of workplace travel planning in urban conurbations? Examples from London, UK*. Paper presented to the 86th Annual Meeting, Transportation Research Board, Paper 07-2164, Session 553: New International Research in Transportation Demand and Parking Management, Washington DC, 21–25 January.

Ubbels, B., Enoch, M. P., Potter, S. and Nijkamp, P. 2004. *Unfare Solutions*, (London: Spon).

UK BIDs (National BIDs Advisory Service). 2011. Home of the national BIDs advisory service. [Online] Available at http://www.bidsuk.org [accessed 4 April 2011].

Wolf, J. F. 2006. Urban governance and business improvement districts: the Washington DC BIDs, *International Journal of Public Administration*, 29, 53–75.

Zuehlke, K. and Guensler, R. 2007. Employer perceptions and implementation of commute alternatives strategies, *Journal of Public Transportation*, 10(4), 171–194.

Chapter 5

Structure: Re-orienting Travel Plans to Increase Effectiveness and Levels of Take-up

1 Travel Plans within Organisations

In its current guise, the Travel Plan is a potentially effective transport planning tool, but one that remains a marginal one. In the opinion of the author, this frustrating position has occurred because one fundamental element of the concept – that is the introduction of a new transport policy delivery agent that is from outside of the transport policy and planning institutional framework – has not been sufficiently recognised. Indeed, one could argue that this 'structural' component of Travel Plan development has been all but ignored.

Accordingly, the aim of this chapter is to address two core inter-related questions:

- Why are Travel Plans not implemented in more organisations?
- Even where organisations have adopted a Travel Plan, why are so many of them almost ignored?

The chapter does this by first revisiting the definition of the Travel Plan and then briefly reviewing the literature on barriers to Travel Plans, which leads onto a series of basic questions around the concept as currently understood, namely:

- What factors currently motivate organisations to undertake Travel Planning – that is WHY should they do it?
- WHAT does doing a Travel Plan mean for an organisation?
- HOW are organisations firstly informed about Travel Plans, and secondly how are they then encouraged and supported to do a Travel Plan?
- WHEN are organisations approached to take up Travel Plans?
- WHO is involved in firstly informing organisations about Travel Plans, and secondly in encouraging and then supporting them in implementing and operating a Travel Plan?

Each of these issues is then explored – first from the perspective of the 'traditional' Travel Plan model, and secondly from a slightly different perspective, that is as a re-oriented model of Travel Plans as perhaps they should be. Finally, conclusions are drawn.

2 The Problems with Travel Plans

As already noted towards the end of Chapter 2, the Travel Plan literature (Bradshaw et al (1997), Coleman (2000), Rye (2002), Cairns et al (2004), Roby (2010), Enoch and Ison (2008) and Enoch and Zhang (2008)) suggests that several issues emerge as barriers to Travel Plans. These include issues resulting from problems caused by a lack of strategic policy direction by government and problems with the Travel Plan industry as a whole, as well as more structural financial and staff resource concerns within organisations themselves (see Table 5.1).

Table 5.1 Barriers to Travel Plan adoption

Level	Barriers to Travel Plan adoption
Government	Lack of vision and leadership by government and local authorities; Lack of integration with wider public policy areas; Personal taxation and commuting issues; Lack of regulatory requirements for Travel Plans.
Travel Plan Industry	Lack of examples due to novelty of the concept; Lack of evidence of how successful Travel Plans are; Lack of public transport operator involvement; Lack of supplier interest in Travel Plans.
Organisation	Companies' self-interest and internal organisational barriers; Non-core nature of Travel Plans; Lack of financial resource within companies; Lack of staff resource (including skilled Travel Plan coordinators); Lack of integration within the wider business.

In addressing these, this chapter will consider the organisational barriers referred to in the table which actually relate to the nature of the Travel Plan concept itself, and crucially to how it is then understood and applied in practice, while the means of overcoming the government and Travel Plan industry barriers – that is the 'support' measures – will be looked at in Chapter 6.

Returning to the common themes that seem to underpin the Travel Plan as a concept (as set out in Table 2.2), these are:

- Travel Plans are not really an instrument themselves, but a delivery mechanism or strategy for other mostly transport-focused measures.
- Travel Plans are delivered by an additional 'agent' that is not a part of the 'traditional' transport policy institutional structure.
- Travel Plans are initiated in two ways: either by the organisation or by government.

- Travel Plans seek to deliver transport and related benefits to the community as well as some more direct organisational benefits to the participating 'delivery agents'.
- Travel Plans are to some extent 'site-specific', that is tailored to the specific contextual circumstances.
- Travel Plans to some extent deliver a package or a strategy of a 'wide variety' of transport instruments.

Overall, then, a catch all definition that emerges from these points is that Travel Plans are a mechanism for delivering a package of transport measures targeted at a specific site by an agent with a strong relationship with the local transport users to deliver transport and wider goals to the organisation and society as a whole.

From this understanding, coupled with the organisational barriers or 'application issues' noted in Table 6.1 (such as companies' self-interest and internal organisational barriers, the non-core nature of Travel Plans, and the lack of integration within the wider business; together with a lack of financial or staff resource within companies), a number of implications arise. These can be framed as the basic questions set out at the beginning of this chapter, specifically:

- What factors currently motivate organisations to undertake Travel Planning – that is WHY should they do it?
- WHAT does doing a Travel Plan mean for an organisation?
- HOW are organisations firstly informed about Travel Plans, and secondly how are they then encouraged and supported to do a Travel Plan?
- WHEN are organisations approached to take up Travel Plans?
- WHO is involved in firstly informing organisations about Travel Plans, and secondly in encouraging and then supporting them in implementing and operating a Travel Plan?

3 The Traditional Travel Plan Model

This section sets out how Travel Plans currently operate from the organisational perspective.

3.1 Motivations

As they stand, Travel Plans are typically initiated either voluntarily by organisations with some form of transport problem, or else by local authorities requiring organisations to adopt a Travel Plan so as to fulfil a planning or other regulatory obligation to help mitigate the travel generation impacts of a new development on the wider community. In both cases, Travel Plans are often not fully integrated into the wider organisational processes (finance, human resources, CSR, environment) as effectively as they might be, and instead they tend to be treated as an estate management or facilities function. Additionally, in the latter case organisations typically treat Travel Plans as a burden imposed upon them to be discharged as quickly and as cheaply as possible rather than as a positive process that might be of benefit to them. Indeed, the expectation is that apparently mandatory Travel Plan requirements are typically almost never enforced and, as a consequence, significant proportions are not as effective as they might otherwise be.

This is unfortunate given the wide range of benefits that an appropriately designed Travel Plan can confer on an organisation, and is largely due to a lack of consideration of how best to communicate with a new type of actor in the transport system that generally has very different motivations and ways of working to pre-existing stakeholders.

3.2 What is Involved?

Introducing a Travel Plan is often perceived to be a significant, even an intimidating step for an organisation to take. This is because of the idea that a Travel Plan is a 'package or a strategy of a wide variety of transport instruments' with corresponding resource requirements, and because Travel Plans seem to require organisations to become transport providers. In addition there are usually expectations of community benefits (and the monitoring of these) by local authorities. Worse, the required activities do not usually explicitly map onto the 'standard tasks' of that organisation while the expected benefits to the organisation itself are not usually seen as being sufficiently attractive to justify such a high level of commitment. For instance, commuting in the UK is not a responsibility of the employer, and yet the majority of Travel Plans seem to primarily focus on home to work travel rather than on business travel (travel in the course of work), which can be a major direct cost to an organisation. Similarly, freight supply and delivery are often treated separately to the Travel Plan.

In other words, Travel Plans are stymied by being an inappropriate product for most organisations in the market place. Specifically, traditional Travel Plans (mostly):

- Only meet the needs of the organisation if it has a transport-related problem – that is a small proportion;
- Require significant resources to be contributed;
- Are not seen as being relevant to the organisation.

From this, it would seem obvious that these barriers would need to be overcome if the Travel Plan concept is to become more mainstream.

3.3 How are Organisations Informed and Encouraged to Participate?

Next, available information has tended to focus on the impact of Travel Plans on transport and related issues – reducing congestion, addressing parking problems, cutting carbon emissions – with rather less knowledge provided on issues such as links to employee health, potential productivity savings, recruitment and retention benefits and direct financial outcomes. Meanwhile, information and encouragement regarding Travel Plans is generally provided by national/regional government and local authority transport departments, and so is often written in a form that is alien to the organisations it is provided for, while the primary Travel Plan advisors are transport consultants. Furthermore, the primary incentives for organisations to participate are regulatory 'sticks' whilst other (normally public sector) organisations participate through directives being issued from senior management. Rather fewer organisations are encouraged to participate in travel planning activities through awareness campaigns, subsidies and/or tax incentives.

Predictably, therefore, the inclination for organisations is to avoid the idea of a Travel Plan for as long as possible, rather than to try out some activity because they were encouraged to do so.

3.4 When is this Done?

Travel Plans are typically 'sold' to organisations as and when resources within a transport authority's budget allow, rather than at specific points in the development of an organisation (the exception here is where organisations requiring planning permission may well be asked to prepare a Travel Plan).

This can be a major sticking point, and one that is starting to be recognised by researchers of individual travel behaviour (for example Scheiner 2007). Essentially, organisations, like people, are more susceptible to messages encouraging them to change their *modus operandi* at certain times of the year, and at particular points in their development. Somehow the promoters of Travel Plans need to factor this point into the marketing of the concept.

3.5 Which Organisations are Involved in the Process?

Travel Plans are predominantly promoted by national/regional government and local authority transport departments, while the primary Travel Plan advisors are transport consultants. Although it is true that some organisations do have relationships with local authority transport departments and/or transport consultants who may have delivered parking plans, integrated transport plans or transport assessments for example, one would suggest that many other organisations that do not perceive a transport problem exists (a sizeable majority?) have not.

Once again, as with individuals, organisations tend to be more accepting of new ideas that challenge their existing outlook and practices from people and organisations that they trust, and these tend to be people/organisations with whom they already have a pre-existing relationship. Consequently, one might suggest that other forms of consultant with more frequent and regularised relationships may be better placed to integrate Travel Plan-related measures into an organisation. So, for instance a tax consultant could probably recommend an organisation adopt a salary sacrifice discounted public transport season ticket for its employees during the most appropriate stage of the budgetary cycle more effectively than a transport consultant. Similarly, a recruitment and retention consultant would likely be more trusted about measures aimed at keeping employees happy – *even if those measures were transport-related*, for example encouraging home or flexible working practices.

3.6 Emerging Threads

Before looking in detail at what this means for the future, it is important to recognise that this traditional model of Travel Planning can work effectively in specific circumstances and still does have a role to play. Accordingly it is assumed that any new Travel Plan model would not necessarily wholly replace the traditional model, although over time one might envisage that such examples would become proportionately less common.

Moving on to pull the themes of the traditional model together, the core issue is to do with whether a Travel Plan is viewed either as:

- a local authority transport policy solution which primarily delivers benefits to the community whilst businesses participate either because they have specific transport-related issues or to improve their corporate image by having 'done the right thing'; or
- a business management tool that meets the needs of that organisation first and delivers transport benefits as a by-product.

In the opinion of the author, it is the marked imbalance between these two positions that is one further reason that progress towards the widespread adoption of Travel Plans has been hampered. Bluntly, this is because it is the first approach that is currently so dominant, a point also raised in Comsis (2002: I–5) which states: "implementation at the employer level as carried through by public agencies focuses almost entirely on public – agency derived goals". As a result, at the moment there would appear to be two current markets for Travel Plans, both of which stem from the local authority 'transport solution' vision. These are:

- Travel Plans required mandatorily by public authorities through planning or other regulations. This is an increasing market but is largely seen (though not exclusively) in a negative light by private organisations, who typically try and avoid their obligations as far as possible.
- Travel Plans adopted voluntarily by organisations so as to address 'traditional' Travel Plan concerns (poor site access, insufficient parking on site, local traffic congestion). This is a niche and almost static market – because one would suspect that many, if not most organisations in this situation already have some form of Travel Plan.

Assuming this to be true, it is perhaps therefore unsurprising that Travel Plans have not progressed beyond their current marginal status.

However, there is another Travel Plan market that could be developed. This assumes the position that most organisations see transport as being only a minor concern of theirs, and instead see teaching children, generating profit, or treating patients as being more of a priority. This approach seeks instead to develop tools that may help with these functions, while delivering transport benefits almost accidentally. For example, several schools in London use the example of how children travel to school each day to educate students about issues ranging from geography to the environment to mathematics – and make use of 'pester power' to influence parents into letting the children walk, cycle or use the bus to school instead of the car. Meanwhile building new retail space on the site of a former car park can help generate profit to an organisation, while hospitals can see increased levels of people walking and cycling as contributing to a healthier society. In other words, the approach contends that if Travel Plans are to be taken up more widely by organisations then a new (possibly additional) 'product' needs to be developed.

In proposing this, it builds on the work of Comsis (2002) referred to earlier and to Roby (2010: 8) which emphasises the need for Travel Plans "to develop from the original concept of influencing travel demand and encouraging more staff to travel in more sustainable ways [that is a transport planning solution] … to become much more … of a business management tool". Encouragingly, the recently released British Standard into Travel Plans largely concurs, in its definition at least, stating that a Travel Plan is "a long-term management strategy for an organisation and its various sites or business park that seeks to deliver transport objectives through positive action and is articulated by a document that is regularly reviewed" (BSI 2008: 1).

Such revelations are not new and indeed are fairly obvious. Yet with few exceptions – for example with the Corporate and Enterprise schemes run by Transport for London and the recent advice offered in the PAS 500 (*op cit*) – the implications appear to have been largely ignored.

4 Re-orienting the Travel Plan Model

This section seeks to present how a new 're-oriented' model for Travel Plans might be constructed, and draws on examples from current practice to provide as vivid an illustration as possible as to how Travel Plans could potentially be developed in the future.

4.1 Motivations

Travel Plans should be re-oriented from being seemingly developed to primarily meet local authority transport and related goals towards primarily meeting the goals of each individual organisation (making sure that the local authority transport goals are still met but almost as a by-product). In practice this means that promoters of Travel Plans need to highlight the cost saving, recruitment and retention, Corporate Social Responsibility and environmental image enhancing benefits of the tools *to the organisation*, rather than pushing the wider benefits to society as often happens now.

Such a perspective is strongly supported in Comsis (2002: III–6), which reports that employers resent regulatory imperatives and success being solely measured "against regional trip reduction goals for air quality improvement or congestion mitigation". Instead, it finds that:

> the prevalence of internal motivation [that is provision of employee benefit, resolution of site constraints, easing of on-site congestion, enhanced employee recruitment, enhanced public image] indicates that employers have not, by and large, been convinced that they should participate in TDM to advance community goals. *Most are not and might never be interested in them* [emphasis added].

Similarly, TCRP (2010: 87) sees the voluntary approach as being more likely to lead to a 'productive' Travel Plan due to them being implemented for reasons of self-interest. It continues:

> It is generally difficult to persuade employers or institutions to adopt [Travel Plan measures] … , *unless they see a clear benefit to their bottom line*. Hence, the existence of a regulatory requirement is no guarantee that this class of measures will be widely adopted … [Experience suggests] that while a number of 'success stories' may emanate from areas with regulatory requirements, almost as many of the successful examples come from areas where there is no formal motivation *beyond economics, concern for employees, or peer pressure* [emphasis added in both instances].

In adopting this viewpoint, there are numerous examples of organisations that have adopted Travel Plans for reasons of economics, concern for employees or peer pressure.

For instance, exclusive car maker Bentley Motors sought to cut costs to both staff and the company while simultaneously enhancing its CSR and environmental image by achieving ISO 14001 accreditation and significantly reducing carbon dioxide emissions (NBTN 2010). To do this, it has integrated environmental factors into its business decisions and targeted both commuting and business trips. Specifically, for commuting it promotes walking, cycling and lift sharing (carpooling). Business travellers must pre-request trips, are only granted authorisation if virtual options will not suffice, and must consider public transport before being able to use the car or fly. Consequently, the company reduced carbon dioxide emissions from commuting by 490 tonnes (or 2.4 million kilometres) in 2009 and saved some £GBP1.5 million (€1.7 million) from business travel in 2008.

Meanwhile at Arlanda Airport's Cargo City outside of Stockholm, Sweden, a Local Travel Plan Network was established as a response to an emissions cap being enacted which threatens to limit the level of carbon dioxide emissions produced by the airport, and hence the potential profitability of the airport (TRAVEL PLAN PLUS 2011). And on similar lines, there could be scope for Travel Plans to be incorporated as part of the Eco-Management and Audit Scheme (EMAS), the European Union's voluntary environmental management programme for organisations, for example.

Next, when the Defence Science and Technology Laboratory consolidated its sites from across the south of England to a single location at Porton Down in Wiltshire, a network of work bus services were introduced primarily to ensure that skilled employees were not lost by them being prevented from getting to work. Moreover, significant efforts were spent on ensuring that travellers could work on the bus – and crucially in having the organisation recognise time spent on the bus as 'work time' – and hence not be penalised by having to travel further than previously (Hall, D. 2009. Personal communication, Travel Plan Officer, Defence Science and Technology Laboratory, Porton Down, 7 July).

As seen in Chapter 3 in the cases of schools and universities, peer pressure between organisations competing through specific competitions or through league tables can be extremely effective in pushing Travel Plans. Regarding other examples, bodies such as ACT Travelwise, MODESHIFT and the National Business Travel Network both run popular Travel Plan award schemes in the UK, as does the Association for Commuter Travel in the USA and the EC-EACI-IEE STEER-funded COMMERCE project at the European level. Interestingly, peer pressure can also be assuaged though accreditation processes. Thus, the PAS 500 set by the British Standards Institute grades Travel Plans as being of bronze, silver or gold standard according to their assessed quality (BSI 2008). Overall, though, while such techniques are obviously working to some degree, there is certainly scope to apply these motivations in a way that would make Travel Plans more appealing to organisations that have not thought about adopting them before.

Perhaps most obviously, developers of league tables and awards (media organisations, campaign groups) and accreditation bodies (professional institutions, standards agencies) in a far more diverse array of sectors (for example management, CSR, environment, finance, facilities management) could be approached and persuaded to incorporate Travel Plan elements into their respective criteria. Examples of this type of activity appear to be rare, though one notable exception is that Travel Plans do feature as part of the Building Research Establishment Environmental Assessment Method (BREAAM) as part of the energy section to minimise CO_2 emissions from transport to and from a building, and which also considers:

- The location of the development;
- Parking and cyclist facilities; and
- Access to public transport and local amenities;

Moreover, this has led to the Stockholm Royal Seaport development of 10,000 apartments and 30,000 jobs in Sweden, introducing Travel Plans in a bid to secure an 'excellent' rating under the BREAAM standard for the area.

In other words, while internal motivations are obviously important to organisations, it is also true that external motivations that are not necessarily directly related to Travel Plans can provide an impetus for Travel Plan adoption.

4.2 What is Involved?

Travel Plan tools should move away from comprehensive and holistic (but potentially intimidating) packages towards being more targeted treatments for specific organisational problems. Hence, it should be possible for organisations to select individual tools 'off-the-shelf' that are simple, cheap and easy to implement, yet are effective at addressing the specific problem identified while delivering transport benefits too. Such off-the-shelf tools could, for example, include pool bike packages (to cut down on short distance business journeys); tax efficiency packages (comprising public transport season ticket discount schemes and employer 'cycle for work' schemes); flexitime recruitment and retention schemes; information highlighting the real cost of parking to employers and proposing cost effective alternatives; and measures aimed at enhancing environmental or CSR performance through measures that happen to reduce car use. In addition, the ideal would be for such measures to become fully integrated within the appropriate organisational management, human resource, tax, environmental management or CSR systems.

In the Netherlands, car leasing firm Athlon offers employers a range of tools to help with their mobility needs (Athlon 2011). These include providing a NS Business Card (a discounted rail card) to lease car drivers so that they can use the train instead of the car if it is more convenient to them; company car sharing (car club) schemes; and also a mobility secretary, who is available round the clock to deliver up-to-date individually tailored travel and traffic information. Similarly, mobility services provider Mobility Mixx offers the rail card/lease car/information elements, and provides them on a single card that can also be used to use the OV-fiets (Dutch Railway's national shared bicycle scheme), park-and-ride services and rental cars, and enables the electronic processing of mileage claims and the management of personal mobility budgets (Mobility Mixx 2011). Also in the Netherlands, Vipre offers staff to man commuter centres and operates vanpool programmes.

In the Province of Quebec, Canada, the merged municipalities of St Laurent and Montreal in 2004 together re-launched its economic development services by equipping itself with a not-for-profit organisation with a mandate to foster a dynamic economy (DESTL 2011). Développement économique Saint-Laurent (DESTL) therefore offers employers a whole range of business services, and these happen to include a series of transport-related tools, for example support with:

- Drafting transportation plans tailored to the business;
- Organising and managing alternative transportation services;
- Promoting alternative means of transportation;
- Presenting businesses' interests to municipal bodies;
- Participation in think-tanks for alternative transportation modes;
- Participation in major events related to the promotion of active transportation;
- The rolling out committees between companies;
- The selection and implementation of resources and measures;
- Budgeting and identifying financial partners;
- The organisation of on-site events;
- Devising a commuting management plan prior to the undertaking of construction projects.

Working practices too can heavily influence travel behaviour and are particularly important where re-oriented Travel Plans are concerned (see Chapter 2), due to their being a core concern to any organisation. So, for example, not-for-profit initiative Work Wise UK (WWUK 2010) seeks to facilitate businesses to 'work smarter' by providing practical information and support to help them introduce:

- Flexible working hours – such as allowing staff to arrive and/or leave at different times. The WWUK website posits that if adopted by enough companies, flexible working would 'stagger' the rush hour spreading it out and making it less intense, reducing road congestion and overcrowding on public transport, reducing travel time and making it a less stressful part of the day for everyone.
- Flexible working locations – such as encouraging staff to work from more convenient locations such as from home or whilst on the move. Here, WWUK (2011) suggests productivity would be improved, through reduced travel times and improved quality of life. Indeed, it adds that providing access to satellite offices, hot desks or other facilities more strategically located could reduce travelling distance and time still further.

Looking at schools, the 'Big Wow' campaign based in London has developed a pack of materials for teachers to encourage their pupils to walk to school. This incorporates rewards (pupils receive a badge if they walk to school once a week for a month); together with a series of lesson plans which help teachers tie the resources with elements of the National Curriculum for geography, information, communications and technology (ICT), history, physical education, English and art and design (Living Streets 2011).

And for residential Travel Planners looking for suitable tools to use, one approach being developed in Helsinki, Finland, is to provide an additional information module to on-line apartment search portals that enable prospective residents to see how accessible their likely destinations (workplaces, children's schools, shops) would be by a range of different modes (Laine 2010).

Related to this, one policy idea that emerged from Cuba in response to a lack of transport supply was to facilitate people who lived far from work to either swap their jobs or their homes with people in similar circumstances (Enoch et al 2004). While implemented by the National Government in the Cuban case, it might be equally possible to establish such a scheme either commercially or else within large organisations, such as for hospital staff, academic staff, teachers and so on. In practice, a spin off from the idea behind this approach has already been applied by Bedfordshire Police in the UK, which has carried out a long-running poster campaign at commuter rail stations in the County targeting police officers commuting to London and suggesting that they change their employer.

Finally, EST (2005) provides a useful summary of several information technology tools that may be useful to the sector in the future (see Table 5.2).

Table 5.2 Opportunities for ICT Travel Plan tools

Travel Plan measure	ICT need
Encouragement for home working.	Remote access to telephone, email and files; 'extranets', training and support.
Tele-conferencing; video-conferencing systems.	Improved public transport information, Intranet and email information services, mapping and route planning tools.
Lift sharing.	Bespoke lift sharing software, intranet bulletin boards, staff travel survey, online survey software, databases.
Publicity and awareness raising campaigns.	Graphic design, mapping services, contact databases.
Planning new bus services.	Mapping and geographic information system (GIS) tools, accessibility software.
Fleet management.	Telematics, fuel management software, routing and scheduling systems.

Source: EST (2005)

4.3 How are Organisations Informed and Encouraged to Participate?

Government and the Travel Plan industry need to do more to inform, support and incentivise organisations to take up Travel Plans. This can be done through re-orienting awareness campaigns to be organisation-focused; by establishing subsidy and tax incentive regimes that reward organisations with Travel Plans; through providing expert advice; and by establishing a clear policy vision that sets out exactly how Travel Plans fit in transport, planning, environmental and fiscal policy areas.

TCRP (2003) points out that marketing material needs to be professionally designed, be updated regularly and contain local reference information if they are to appeal to a business audience. Transport for London is one body that has taken this message on board and has worked hard to produce materials specifically targeted to a business audience (see TfL 2010). Likewise, Heather McInroy, the director of the National Business Travel Network sees the importance of Travel Plans being sold to organisations on business lines (McInroy, H. 2011, Personal communication, Director, National Business Travel Network, Amsterdam, 25 February), and has modified the content and presentation of the materials produced, and the engagement strategy of her organisation accordingly (see NBTN 2011).

Perhaps more directly, in Cardiff, Wales, more than 40 organisations have benefitted from grants of up to £GBP10,000 (€11,400) for physical and soft measures to support sustainable commuting which has led to a number of new and improved Travel Plans in the city. Additionally, support in the form of problem solving and resource sharing activities is provided by two local Travel Plan networks – one for higher and further education organisations and one for other companies – which together comprise of 20 members.

And in the Commonwealth of Virginia, organisations can receive up to $USD35,000 (€24,400) ($USD50,000 or €34,900 in some areas of Northern Virginia) of public money to offset the start-up or expansion costs of a telework programme if they apply to Telework!VA (Telework VA 2011). In addition, a suite of e-learning materials is available to support organisations in setting up their scheme.

However, providing information and encouragement to organisations adopting Travel Plans is one area where transport authorities need to develop not only a proactive approach, but also their reactive capabilities. As will become clear in the next section, this is because organisations are more susceptible to new ideas and to change at certain times in their development, and so it is vital that Travel Planners are able to exploit these opportunities as and when they arise.

4.4 When is this Done?

Regarding the timing issue, a far more responsive arrangement is needed whereby organisations should be able to tap into Travel Plan activities at their convenience – that is when they have a problem that needs addressing or when their circumstances significantly change – not as now where the local authority decides to act in a particular neighbourhood due to a transport problem being identified there or because resources have suddenly become available. For example, particularly appropriate instances within an organisation could be: during departmental re-structuring exercises; instances when staff terms and conditions are being re-negotiated; or during fiscal budgeting exercises. By contrast TCRP (2003) notes that particularly bad times are when employers are facing financial problems or mergers and when employers have key positions unfilled (for example, a human resources or benefits manager).

The Cambridgeshire Travel to Work Partnership is one organisation that has adopted one such approach. Here, Travel Plan officers routinely read through local business journals and participate in various local business networks to determine which organisations are about to re-locate, and these are then approached by the Partnership and asked if they would be interested in developing a Travel Plan (Webb, M. 2011. Personal communication, Manager, Travel for Work Partnership, Cambridgeshire County Council, Amsterdam, 24 February). In the USA too, the Arlington Economic Development in Virginia interacts with prospective employers about transport issues whilst they are in the process of making their decision as to where to physically locate their organisation (Jennings, H. 2011. Personal communication, Manager, Arlington Economic Development, VA, 24 January)

Another example of timed interventions occurs in Sweden, where Dr Markus Robert – the founder of the CERO model – targets large companies that promote themselves as being CSR-conscious. He then attends various CSR network events where such issues are discussed and 'broadens the scope' of how CSR is perceived to include commuting and business travel, before selling CERO as a means of helping companies save money, reduce environmental impacts and achieve other indirect benefits (employee health, for example) (Roberts, M. 2011 Personal communication, Director, CERO, Stockholm, 25 February). Crucially, from the timing perspective, organisations that have staff that are attending meetings about CSR are likely to be open to new ideas about how they might enhance their performance against CSR benchmarks.

Meanwhile, external events can also provide opportunities for Travel Plan interventions. One high profile example of this is the Olympic Games set for London in 2012, where the Olympic Development Authority is already engaged with businesses to 'maintain business continuity' throughout the period of the event. ODA (2010) notes that the Olympic and Paralympic Games will take place in the summer in London and across the UK, and that they will impact on businesses in four main areas:

- journeys to and from work for you and your staff;
- business travel;
- travel for your customers or other visitors; and
- deliveries/collections and other suppliers.

Consequently, it suggests that organisations:

- Firstly reduce non-essential journeys by encouraging and managing annual leave; working from home; temporarily relocating staff to other locations as appropriate; using alternative meeting methods such as conference calls or videoconferencing; rescheduling meetings; offering alternative supply methods; or stockpiling goods prior to the Games.
- Secondly, if travel cannot be avoided then manage essential journeys by scheduling trips to avoid busy periods; re-routing trips to avoid busy locations; using less busy methods of travel; altering working/opening hours; promoting alternate outlets; sharing and consolidating deliveries; and; receiving and collecting goods at less busy times.

To implement these ideas, it proposed that organisations develop an Action Plan for the Games, and subsequently test and communicate the plan amongst staff.

Such a situation, when transport conditions will be dramatically different for everyone in most parts of London, could seemingly offer a unique chance to persuade organisations of the benefits of 'doing transport differently' on a more permanent basis.

On similar lines, but rather more commonplace, is the use of road works to stimulate a 'conversation' with the motorist about their travel behaviour to determine how else they might satisfy their travel needs (see Chapter 3 for more on this).

Of course relocation is the one timing point (in the UK at least) that is already targeted by the Travel Plan process, in that Travel Plans are often required as a condition of planning permission being granted. Unfortunately, while theoretically this provides a useful opportunity to implement a useful Travel Plan, in practice the most influential decision from a transport perspective – that is where the new development is actually to be located – has almost always already been made. This means that the Travel Plan is not a pro-active tool that could design out major issues before they arose, but it is generally a 'sticking plaster' attempting to make the best of a sub-optimal solution. Instead, far more effective would be for the Travel Plan concept to be incorporated in the relocation process long before, that is prior to the local transport authority being aware of it.

4.5 Which Organisations are Involved in the Process?

From this, it is obvious that 'getting the timing right' means working with the right agencies, at the most basic level as a means of 'getting a foot in the door'. More fundamentally, Travel Plans theoretically work well from a local authority transport policy perspective because they deliver transport improvements at the most appropriate delivery level. Unfortunately, one reason there are not more Travel Plans is because a local authority transport officer or transport planning consultant is probably not always the most appropriate 'messenger' to communicate the concept to the potential implementing organisation. As noted earlier, although it is true that many organisations already have a relationship with local authority transport departments, it is likely most do not. Rather, Travel Plan tools maybe need to be 'sold' by bodies that already enjoy an established relationship with organisations, for example management consultants, tax advisers, CSR gurus and environmental consultants instead. One further advantage of this approach is that identifying the 'right' person in any organisation to contact can be extremely difficult (TCRP 2003).

Consequently, it might be suggested that these alternative forms of consultant with more frequent and regularised relationships may be better placed to integrate Travel Plan-related measures into an organisation. So, for instance a tax consultant could probably recommend an organisation adopt a *Transitchek*-style salary sacrifice discounted public transport season ticket for its employees more effectively than a transport consultant, particularly if proposed during the most appropriate stage of the budgetary cycle. Similarly, a recruitment and retention consultant would probably be more able to introduce measures aimed at keeping employees happy (even if those measures were transport-related such as encouraging home or flexible working practices), especially if already engaged by the organisation to look at such issues.

Of course it is clear that there will be some Travel Plan 'measures' that can only be introduced by transport consultants, and the author is not proposing that transport consultants have no role to play. However, one key conclusion must be that transport consultants might not be best placed to mainstream Travel Plan measures in this proposed re-oriented form. Instead, to really take advantage of the untapped opportunities afforded by providing Travel Plan 'tools' to organisations with no apparent transport problem as a business process enhancement, then transport consultants should be looking towards creating partnerships with these other forms of consultant. Such an arrangement would see transport consultants generating a wider range of client services (namely Travel Plan measures) for their partner consultants in return for more frequent and regularised relationships with client organisations.

Meanwhile, a slight variation on this approach is proposed in TCRP (2003), which suggests developing partnerships with business groups such as chambers of commerce and human resource organisations to make links with employers. OECD and ITF (2010) cite the PDE PRO® programme of the Grenoble Chamber of Commerce (CCI) in France in this regard. This began in 2004 as a response by the CCI to worsening traffic affecting productivity and encouraged businesses through various media to conduct a site-specific mobility audit and then an appropriate Travel Plan. As of 2006, when the pilot was due to end, some 2,000 companies of 20 or more employees had been contacted, of which 85 had some involvement and 12–15 had developed Travel Plans. The €120,000 a year costs were covered by the project partners which included local public transport operators and local and regional government agencies.

One good example of where an existing business network was 'hijacked' by the Travel Plan concept is the Better Bankside Business Improvement District on the South Bank of the River Thames in London. Here, European funding through the OPTIMUM2 project enabled a team led by the London Borough of Southwark to meet with the BID managers and convince them that providing Travel Plan measures for their members would be a sensible idea. Subsequently, Travel Planning has become a core activity (see Tyler et al 2007).

In addition, establishing networks of organisations already engaging in Travel Plans in particular neighbourhoods and/or in particular sectors of the economy (such as hospitals, universities, government offices) could generate additional peer-to-peer information and support to further increase the take up (see Chapter 4 for discussion on this).

The National Business Travel Network too, is a strong example of where transport issues have been superimposed on a pre-existing network, in this case as a subgroup of the Business in the Community network of businesses (BITC 2011), which comprises some 850 large organisations and more than 3000 SME 'follower' companies that have an interest in being socially responsible. Similarly, independent consultant Minze Walvius works through networks of facilities managers in the Netherlands and provides 'off-the-shelf' mobility management tools to meet the needs of this group (see earlier reference to Mobility Mixx Travel Plan products). Finally, Travel Planners at the Cambridge Local Travel Plan Network of the TRAVEL PLAN PLUS project latched on to an initiative by the local Primary Care Trust (a healthcare provider) to present health-related Travel Plan alternatives as part of a 'Staying Healthy at Work' road show on the site of the network (Webb 2011).

However, although a simple idea in theory, the reality can be that 'piggybacking' on existing networks is not always simple, either because such networks are not interested in the concept, or else because appropriate networks simply do not exist. So, in the latter Cambridge case for example, sustained efforts by the Travel for Work Partnership to break into a local economic development partnership have so far foundered (Webb 2011). In such situations, though, it is sometimes useful to bear the lessons from the 'timing' section in mind, and either wait for circumstances to change or for a new network to come into being.

4.6 Emerging Threads

Such measures appear to have an influence when applied independently, yet one suspects that any benefits would be magnified if they were to be implemented together. So far as the author is aware, however, such a scenario does not yet exist, although there are some examples of programmes that are moving in that direction.

As alluded to in Chapter 4 and earlier in this Chapter, the Better Bankside Travel Planning Group (BBTPG) meets several of the 're-oriented' Travel Plan criteria. First, the organisations participate through self-interest and not usually through coercion (apart from where new developments require planning permission). Second, the degree of organisational involvement required is limited due to the development of a series of 'off-the-shelf' measures supplied both through Transport for London and through the OPTIMUM2 European Project, and due to the existence of the BBTPG which 'spreads the burden' and which helps apply the measures. Third, the organisations are informed and encouraged to participate through the use of carefully targeted incentives rather than by being compelled to join in by dictat. Fourth, regular meetings of the BBTPG maximise the opportunities for potential members to become involved, though as far as is known there are no specific measures targeting organisations at particular points in their development. Lastly, the key delivery agent is through the management company of the pre-existing Business Improvement District, a body that is already well known about and trusted, and by the peer-to-peer network opportunities that this engenders.

Another example is the case of the CERO model, again already mentioned earlier in the Chapter. Here, after an organisation is targeted and (hopefully) convinced, then outputs and processes are inventoried with respect to the level and benefits of business and commuting travel, and most importantly the direct and indirect costs (Roberts, M. 2011. Personal communication, Director, CERO, Stockholm, 25 February). After this, the results are presented to the senior management at an interactive workshop, and a planned way forward is agreed which importantly aims to adapt the culture of the organisation to properly incorporate the measures proposed. At this point, more specialist mobility management service providers (of company car tax advice, 'clean vehicles' and fleet management services for example) are invited where appropriate. Interestingly, Roberts agrees that timing is crucial in persuading organisations to participate, and has also been considering working in partnership with non-transport bodies as a means of improving the chances of this occurring.

Finally, it is interesting to note that in the USA, many of the features of the re-oriented Travel Plan (such as the organisation focused and 'off-the-shelf' measures being delivered by a specialist Travel Plan sector), have been in place over many years. Moreover, more recently, practice in the Netherlands has shifted towards a similar approach (Metz, F. 2011. Personal communication, Director, KPVV, Amsterdam, 24 February). Such a development in probably the two most 'mature' Travel Plan environments, surely reinforces the case for re-considering how the Travel Plan concept is taken forward in the future.

5 Conclusions and Recommendations

The aim of the chapter was to explore why Travel Plans are not implemented in more organisations and more effectively within those organisations, by examining the current Travel Plan concept and then proposing a new approach to help Travel Plans to fully realise their potential as an effective transport policy tool.

In summary, it finds that Travel Plans currently often very much reflect their local authority heritage as a means of addressing transport problems. This means that they are usually only adopted by organisations with a transport-related problem that needs to be addressed – a niche market – in spite of Travel Plans delivering sometimes significant benefits both to the organisations and to society more generally. The key lesson is that for Travel Plans to be adopted more widely, then more organisations need to be tempted to adopt them, and these view the world rather differently. Thus, for the longed for 'step change' in take up to be delivered then an additional Travel Plan model should be developed (see Table 5.3 for a detailed summary).

Specifically, these new Travel Plans need to focus on the needs of organisations more than those of transport policy makers; should consist of properly targeted individualised measures rather than a holistic 'package'; and should be more attractively incentivised and less coerced. In addition, Travel Plans need to be implemented at the behest of the organisation rather than the local authority; and should be promoted by faces familiar to organisations (for example management consultants, environmental experts) rather than by transport consultants with whom an organisation is not likely to have a pre-existing relationship.

Table 5.3 Model features compared, and possible implications

Features	Traditional Travel Plan	Re-oriented model	Implications
Overall description	Mechanism for implementing site-specific transport planning tools through traffic generating organisations to solve local transport problems whilst providing benefits to organisations.	Range of 'off-the-shelf' organisational tools designed to meet organisational goals (for example cost reduction, recruitment and retention, CSR, environmental) whilst addressing local transport problems.	Shift from a minority of organisations being pseudo transport planning centres to sustainable transport objectives being delivered by stealth as a by-product of everyday process improvements.
Motivation for participation	Planning permission or to meet environment/CSR inspired public policy directives (mandatory). Site-based transport issues (such as parking, congestion, access), environmental/CSR credibility or cost reduction (voluntary).	Organisational self interest principally economic benefit, to improve employee conditions, or as a result of peer pressure.	Shift from mandatory and transport-related motivations towards broader organisational ones. Less need to monitor compliance.
Degree of organisational involvement required	Often significant due to emphases on 'package' of measures and 'site-specific' modifications. Transport not core to organisation so often management issues.	Limited due to tools being introduced in tandem with broader organisational changes.	Sustainable transport objectives would be met as a result of broader changes thus no longer requiring additional resources.
Available information and support	Information public policy rather than organisation facing. Support often coercive.	Strongly focused on organisation interests – costs and benefits. Support incentive-based.	Information tailoring needed to focus on organisational needs. Shift needed from enforcement to inducement regime. More resources needed to work well.
Timing	Travel Plans pushed when resources allow or when organisations require planning permission. Little account taken of organisational context.	Tools proposed when management, CSR, environmental, recruitment and retention or financial systems under review.	Shift from pro-active resource-led stance by transport authorities to reactive organisation-led regime.

Table 5.3 *Continued*

Features	Traditional Travel Plan	Re-oriented model	Implications
Key supporting stakeholders	Local authority transport officials and transport consultants dominant.	Management, CSR, environment firms in partnership with transport consultants. Such firms have access to organisations at the crucial timing points noted above and often have pre-existing relationships.	Current Travel Plan industry is immature, with few actors involved and limited products in place. Transport consultants should build strategic alliances with broader-based firms to extend potential market.
Performance characteristics	When done properly can achieve significant site-level transport improvements. Network level impacts less impressive due to low take up among organisations.	Expected site-by-site transport impacts may fall but not necessarily – measures would be more aligned with organisational goals and integrated with organisational processes. Network level impacts likely higher through more organisations adopting them.	Re-oriented approach would supplement, not replace, the traditional Travel Plan concept by instilling 'sustainable transport friendly' processes across any organisation that seeks consultancy advice and not just those with transport problems or a need for planning permission.

Table 5.3 summarises how such a re-oriented Travel Plan model might look, and suggests what the implications of such a shift may be.

First, as many of the perceived barriers to take up as possible (companies self-interest and internal organisational barriers, the non-core nature of Travel Plans, and the lack of integration within the wider business; together with a lack of financial or staff resource within companies) are directly addressed so that the re-oriented Travel Plans should actually align with the needs and motivations of the organisations concerned and require far less in terms of 'additional' organisational resource.

Next, the motivations for the organisations switch from being largely reactive to almost wholly pro-active, coupled with the use of expert agents with whom the organisations already have an existing relationship should help lead to the creation of less confrontational and more cooperative (productive?) partnerships between organisations and transport authority. In the same vein, the role of the local transport authority changes from initiator/regulator to facilitator/service provider incentivising participation in Travel Planning rather than penalising non-compliance.

For the organisations, then, the advantages of the re-oriented approach are theoretically manifested in enhanced organisational performance with relatively limited 'non-core' organisational input, delivered at a convenient time and by a trusted intermediary.

Meanwhile for the local authority, although the importance of the public policy objectives is (deliberately and explicitly) played down somewhat, the trade-off is that the likely increased commitment of the organisations and the improved partnership environment hopefully delivers a higher level of benefits than before.

These characteristics are summarised in Table 5.4.

Table 5.4 Resulting characteristics of both Travel Plan models

Traditional Travel Plan characteristics	Characteristics of the re-oriented model
Travel Plans are not really an instrument themselves, but a delivery mechanism or strategy for other mostly transport-focused measures.	'Travel Plans' are not really an instrument themselves, but an application mechanism or strategy for organisation-focused measures that indirectly deliver transport and related benefits.
Travel Plans are delivered by an additional 'agent' that is not a part of the 'traditional' transport policy institutional structure.	Travel Plan measures are applied by an additional 'agent' that is not a part of the 'traditional' transport policy institutional structure.
Travel Plans are initiated in two ways: either by the organisation or by government.	Travel Plan measures are predominantly marketed to organisations by a wide range of service providers, with the support of government agencies.
Travel Plans seek to deliver transport and related benefits to the community as well as some more direct organisational benefits to the participating 'delivery agents'.	Travel Plans seek to deliver explicit organisational benefits to participating 'application agents' and to generate transport and related benefits to the community as a by-product.
Travel Plans are to some extent 'site-specific', that is tailored to the specific contextual circumstances.	Travel Plan measures are organisation and site appropriate.
Travel Plans to some extent deliver a package or a strategy of a 'wide variety' of transport instruments.	Travel Plans apply one or more complementary measures according to the needs of the organisation.

Emerging from the characteristics presented in Table 5.3 is essentially a definition of the re-oriented Travel Plan concept. This is that: "Re-oriented Travel Plans comprise one or more complementary measures applied within an organisation to help address its own goals whilst (indirectly) delivering transport and related benefits to the community".

Interestingly, what this represents is the (last?) stage in the journey first described in Figure 2.1. This starts from the traditional transport planning approach (top left in figure), whereby government seeks to change the travel behaviour of individuals directly; through the traditional Travel Plan approach (top right), where government targets individual travel behaviour via mediating organisations like employers; to a third re-oriented Travel Planning stage (bottom right), where government seeks to persuade the mediating organisations to change the travel behaviour of individuals via a broad range of expert knowledge providers in the management, CSR, human resource, tax and environmental management consultant sectors (see Figure 5.1).

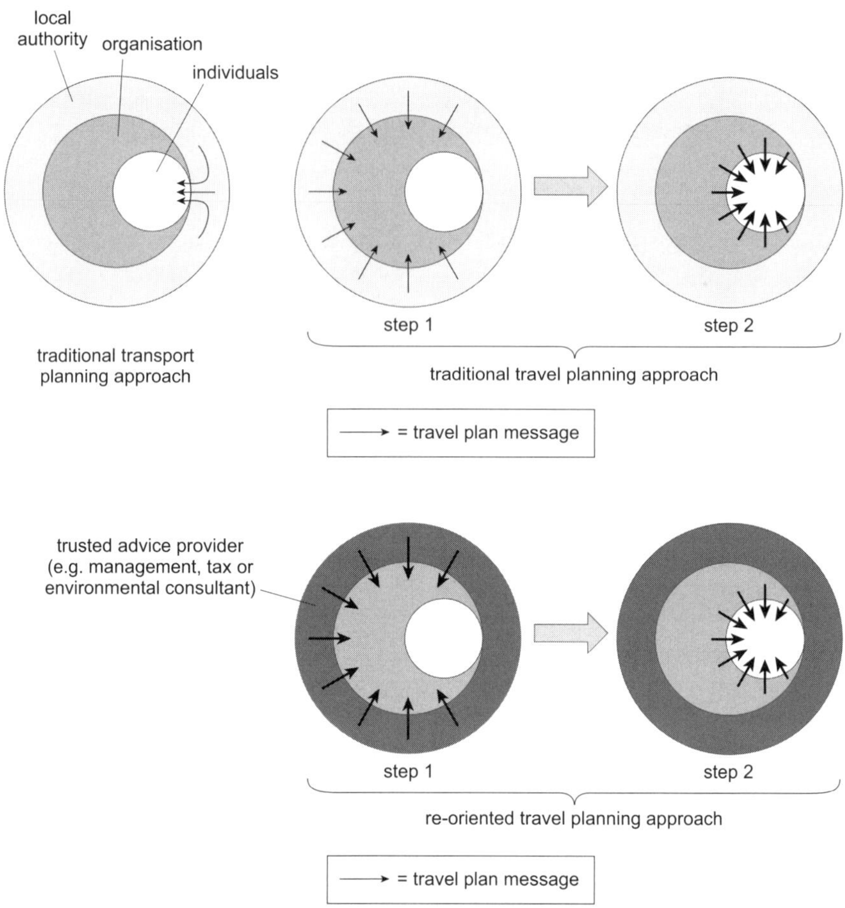

Figure 5.1 From transport planning to the re-oriented Travel Plan

5.1 Applying the Re-oriented Travel Plan

In implementing the new concept, there are several challenges to be faced.

First is scepticism over new ideas and the limited (though growing) evidence as to whether it would work in practice.

Second, of crucial importance would be the development of a vibrant mobility management services sector. So far, beyond a few pioneers outside of the USA and the Netherlands at least, this is not in place, and past efforts suggest that there is a nervousness that such a market cannot exist without a policy environment that is more benign to Travel Planning than currently (see Chapter 6).

Third, and very closely related to the second point, concerns the engagement of the next generation of Travel Plan champions, that is the management, CSR, tax, HR, environmental management consultants. Such a prospect is daunting to say the least given that winning over practitioners even in transport-related sectors like planners, geographers, engineers and public transport operators to any significant degree has been extremely difficult.

Of course, these challenges should not be ignored, and yet there are grounds for thinking that circumstances could change. These can be encapsulated in two key points:

- By definition the re-oriented Travel Plans would be easier, cheaper and quicker for organisations to adopt, and hence should appeal to far more organisations and prove even more attractive for local authorities to support than the traditional model.
- Crucially, re-oriented Travel Plans are not actually a new concept at all, but rather an evolution of the traditional approach which is allowing the changes to happen incrementally and effectively on a scheme by scheme basis. Accordingly, it is likely that over time the traditional Travel Plans will gradually morph along the lines described, hopefully stimulating the mobility management services sector from the 'bottom-up' in a steady and sustainable way which would gradually generate interest amongst the wider target group of stakeholders. Indeed, as already reported throughout this chapter there are several illustrations of cases where elements of the re-oriented model are being successfully and enthusiastically implemented.

Moreover, in the UK context at least, such moves chime with current political philosophy of the so-called Big Society, whereby the role of government is reduced and its place taken by business and community groups. In a speech to the Annual General Meeting of Business in the Community (the parent of the NBTN), UK Prime Minister David Cameron mentioned that business responsibilities ought to include policies for sustainable travel and for more flexible working (Cameron 2010).

Paradoxically, then, it could be argued that the more effective the (re-oriented) Travel Plan becomes the less visible it is. Such an idea – whereby the Travel Plan entity dies but its ideals live on in a more sustainable form – essentially forms a key component in how Travel Planning needs to be reoriented from the inside if the concept is ever to more fully realise its promised but so far undelivered potential as an effective transport planning and wider public policy tool.

In many ways, such an idea of re-orienting the design of Travel Plans towards the needs of the user tallies with views presented in theoretical constructs seeking to advance particular technologies, policies and behaviours, for example Diffusion Theory, Strategic (Policy) Niche Management and Behavioural Economics (see Rogers 2003; Hoogma et al 2002; and Thaler and Sustein 2009). In each case, the aim is to somehow minimise the internal barriers preventing adoption, that is redefine the 'choice architecture' for potential adoptees. Yet this is only half the story and what is also required is the provision of external 'niche protection'. Of course such 'support' measures have already been explored in abstract in the previous chapter, but in the next chapter some of these ideas are taken forward and more specific recommendations are proposed for government and other concerned stakeholders as to how to enable Travel Plans to reach their full potential as a business management and transport policy tool.

Traditional Travel Plans often focus on public policy objectives such as reducing car use with an expectation that appropriate monitoring will be carried out

Bedfordshire Police targeted police officers travelling to work in London in a recruitment poster campaign with the prospect of a shorter and cheaper commute if they switched employer

References

Athlon Car Lease Nederland BV. 2011. *Athlon Sustainable Mobility Plan*, Leaflet, (Almere: Athlon). [Online] Available at http://www.athloncarlease.nl. [accessed 6 April 2011].

Business in the Community. 2011. *Business in the Community Stands for Responsible Business*, BITC. [Online] Available at http://www.bitc.org.uk [accessed 6 April 2011].

Bradshaw, R. 1997. *Employers' Views on Staff Travel Issues: Mobility Strategy Applications in the Community*, Unpublished Report, (London: University of Westminster Transport Studies Group).

British Standards Institute. 2008. *National Specification for Workplace Travel Plans*, Publicly Available Specification 500, (London: British Standards Institute).

Cairns, S., Sloman, L., Newson, C., Anable, J., Kirkbride, A. and Goodwin, P. 2004. *Smarter Choices – Changing the Way we Travel*, Final Research Report, (London: Department for Transport).

Cameron, D. 2010. *Business and the Big Society*. Speech to the Annual General Meeting of Business in the Community, London, 2 December.

Coleman, C. 2000. Green commuter plans and the small employer: an investigation into the attitudes and policy of the small employer towards staff travel and green commuter plans, *Transport Policy*, 7, 139–148.

Comsis Corporation. 2002. *Public Agency Guidance on Employer-based TDM Programs*. Prepared for Transit Cooperative Research Program Project B-4, TCRP Web Document 22, Contractor's Final Reports, December. (Silver Spring: Comsis Corporation).

Développement économique Saint-Laurent. 2011. *All About Employee Mobility,* [Online] Available at http://www.destl.ca. [accessed 5 April 2011].

Energy Savings Trust. 2005. *Travel Plans: New Opportunities for Information Technology Providers*, Transport Energy Best Practice, (London: EST).

Enoch, M. P., Warren, J. P., Valdes Rios, H. and Henriquez Menoyo, E. 2004. The effect of economic restrictions on transport practices in Cuba, *Transport Policy*, 11(1), 67–76.

Enoch, M. P. and Ison, S. G. 2008. *Mainstreaming Travel Plans in the UK: Policy Proposals for Government*, (London: National Business Travel Network and Department for Transport), November. [Online] Available at http://www.nbtn.org.uk [accessed 8 April 2011].

Enoch, M. P. and Zhang, L. 2008. Travel plans, in *The Implementation and Effectiveness of Transport Demand Management Measures: An International Perspective*, edited by S. G. Ison and T. Rye, (Aldershot: Ashgate), 233–256.

Hoogma, R. J. F., Kemp, R., Schot, J. and Truffer, B. 2002. *Experimenting for Sustainable Transport: The Approach of Strategic Niche Management*, (London: Spon Press).

Laine, T. 2010. Harnessing apartment search portals to MM. Paper presented at Moving People Bridging Spaces, *European Conference on Mobility Management*, Session 18, Graz, Austria, 5–7 May.

Living Streets .2011. *The Big Wow*. [Online] Available at http://www.walktoschool.org.uk/bigwow [accessed 11 March 2011].

Mobility Mixx. 2011. *Welcome to Mobility Mixx*. [Online] Available at http://www.mobilitymixx.nl/ [accessed 6 April 2011].

National Business Travel Network. 2010. *Bentley Motors: Case Study*, 7 September. [Online] Available at http://www.nbtn.org.uk [accessed 8 April 2011].

National Business Travel Network. 2011. *Homepage*. [Online] Available at http://www.nbtn.org.uk [accessed 6 April 2011].

Olympic Delivery Authority. 2010. *Keep on Running: Keep your Business Running Smoothly during the 2012 Games*, Leaflet, (London: ODA).

Organisation for Economic Cooperation and Development and the International Transport Forum. 2010. *Effective Transport Policies for Corporate Mobility Management*, Transport Research Centre, (Paris: OECD).

Roby, H. 2010. Workplace travel plans: past, present and future, *Journal of Transport Geography*, 18(1), 23–30.

Rogers, E. M. 2003. *Diffusion of Innovations*, (New York: Free Press).

Rye, T. 2002. Travel plans: do they work? *Transport Policy*, 9(4), 287–298.

Scheiner, J. 2007. Mobility biographies: elements of a biographical theory of travel demand, *Erdkunde*, 61(2), 161–173.

Telework VA. 2011. Telework Virginia, [Online] Available at http://www.teleworkva.org [accessed 6 April 2011].

Thaler, R. H. and Sustein, C. R. 2009. *Nudge*, (London: Penguin).

Transit Cooperative Research Program. 2003. *Strategies for Increasing the Effectiveness of Commuter Benefits Programs*, TCRP Report 87, (Washington DC: Transportation Research Board).

Transit Cooperative Research Program. 2010. Employer and institutional TDM strategies, Chapter 19 of *Traveller Response to Transportation System Changes*, TCRP Report 95, (Washington DC: Transportation Research Board).

Transport for London. 2010. A New Way to Work, transport for London. [Online] Available at http://www.tfl.gov.uk [accessed 18 August 2010].

TRAVEL PLAN PLUS, 2011. *Local Travel Plan Networks: Final Implementation and Evaluation Report*, Work Packages 4 and 5, TRAVEL PLAN PLUS Project IEE-07-592, (Brussels: EC-EACI-IEE-STEER Programme). [Online] Available online at http://www.travelplanplus.eu [accessed 2 May 2011].

Tyler, S., Enoch, M. P. and Zhang, L. 2007. Setting up local travel plan groups: the future of workplace travel planning in urban conurbations? Examples from London, UK. Paper presented to the 86th Annual Meeting, *Transportation Research Board*, Paper 07-2164, Session 553: New International Research in Transportation Demand and Parking Management, Washington DC, 21–25 January.

Workwise UK. 2011. *Working Smarter, Living Better*. [Online] Available at http://www.workwiseuk.org/index.html [accessed 6 April 2011].

Chapter 6
Supporting Travel Plans: What Role for Public Authorities and the Wider Travel Plan Industry?

1 Travel Plans and the Role of Local Authorities

This chapter details the development of the Travel Plan in terms of the role of the public authority and the wider Travel Plan sector. Specifically, it notes how to start with it was the organisations themselves that initiated Travel Plans whilst public authorities played a relatively low key role. However, with the realisation of the potential benefits of the widespread take-up of the instrument, not only public authorities but also other private stakeholders (consultants, service providers, public transport operators) began to become more involved in a variety of roles, providing a variety of forms of support.

Accordingly, this chapter will first identify the stakeholders involved in Travel Planning and discuss the various roles they can play. It then presents the issues in Travel Planning that need addressing, before exploring the various forms of support available in more detail.

2 Stakeholders in Travel Planning and their Roles

Travel Plans are a branch of sustainable transport, and consequently incorporate the same stakeholders, namely:

- National and regional government – sets over-arching policy goals to be met.
- Local government – seeks to apply national and regional policy at the local level.
- Transport suppliers – include public transport and road network operators who directly provide transport services to the users.
- Transport users.
- Interest groups – seek to represent their constituents (for example, transport users, residents, older people, council employees, local businesses, professional institutions) to ensure that their interests are not overlooked in any proposed scheme.
- Transport consultants – provide advice to above groups where and when requested.

In addition though (as noted in Chapter 2), there is the complication of extra stakeholders being involved – the Travel Plan organisations. Not only that, but as the 'delivery agent', these extra stakeholders actually have the most influence on determining how effective the Travel Plan will be. Moreover, such organisations typically have only limited transport knowledge, have very different perspectives on the issues being addressed (and on how to address them), do not usually have any prior experience of how to work with the rest of the stakeholders in the transport sector, and are often (though not always) unwilling and unenthusiastic participants in the process.

As a consequence, such organisations require special treatment if Travel Plans are to be successfully introduced and operated. Specifically, Comsis (2002: III–7) forms three basic premises on which it suggests support from public agencies should follow, namely:

- Public goals must drive agencies' consideration of TDM [Travel Plans], but TDM efforts will be most successful if agencies motivate employers to participate for internal reasons, rather than to solve external, public goals.
- Public agencies that desire employer-based TDM results must visibly and tangibly support employer-based programmes.
- Successful public support requires extensive coordination, communication, and cooperation of public agencies and private groups and clear definition of roles and responsibilities of each.

3 Travel Plan Issues and an Approach to Address Them

As discussed early in Chapter 5, the Travel Plan literature identifies a series of issues that are preventing Travel Plans being as effective as they otherwise might be, and in order to address these divided them into whether they were internal 'structural' issues (which were then discussed later in that Chapter) or external barriers that required action from other stakeholders. Table 6.1 therefore develops this further by first determining whether each issue is a problem to do directly with Travel Plans or with the wider context, second by grouping the issues into similar 'types', and third by allocating them according to the stakeholder to which they most closely relate. Note that the 'Travel Plan industry' here comprises not only the organisations that introduce Travel Plans, but also (and mainly) Travel Plan advocacy groups, Travel Plan consultants, Travel Plan service providers and Travel Plan officers in local authorities.

Table 6.1 Travel Plan issues, actions and actors

Issue	Issue type	Action required	Action required by:
Companies' self-interest and internal organisational barriers.	Internal	Redesign Travel Plan concept to meet organisational needs.	Travel Plan industry.
Taxation issues.	External	Modify fiscal policy.	National government.
Lack of examples due to novelty of the concept.	Mostly internal	Encourage take up through information, incentives and redesigning Travel Plan concept.	Travel Plan industry and government.
Lack of staff resource.	Mostly internal	Redesign Travel Plan concept to meet organisational needs and subsidise.	Travel Plan industry and government.
Lack of financial resource.	Mostly internal	Redesign Travel Plan concept to meet organisational needs and subsidise.	Travel Plan industry and government.
Lack of small- and medium-sized enterprise involvement.	Mostly external	Encourage take up through information, incentives and redesigning Travel Plan concept.	Government and Travel Plan industry.
Lack of public transport operator involvement.	External	Government needs to 'persuade' operators of benefits.	Public Transport operators and government.
Lack of regulatory requirements for Travel Plans.	External	Regulations would increase take up of Travel Plans.	National and local government.
Lack of strategic direction and a lack of leadership in taking Travel Plans forward.	External	Strategic direction would lead to (or at least enable) other direct and indirect supporting actions to be implemented.	National and local government.
Insufficient monitoring and enforcement.	Mostly external	More data/information would address this.	Government and Travel Plan industry.
A feeling that the Travel Plan product is often not fit for purpose.	Mostly internal	Redesign Travel Plan concept to meet organisational needs.	Travel Plan industry and government.
Travel Plan co-ordinators are often junior positions without grounding in transport planning issues.	Mostly internal	Education and training programmes may help here.	Travel Plan industry and government.
A lack of evidence as to the success of Travel Plans making it difficult to convince senior management.	Mostly internal	More data/information would address this.	Travel Plan industry and government.
Wider transport, planning, environmental, social and economic policies providing contradictory signals to transport system users as to how they should behave.	External	Leadership and strategic direction on the role of Travel Plans would help with this.	National and local government.

From Table 6.1 it is evident that several types of measure emerge, specifically: redesign the Travel Plan concept to meet the needs of organisations, provide strategic direction and leadership, modify the tax system, subsidise Travel Plans, provide better information, encourage more involvement from public transport operators, consider introducing regulations, and provide education and training for Travel Planners. Of these, all categories apart from the first (redesigning the Travel Plan concept – which was explored in Chapter 5), will be addressed in the following sections. Before this, however, it is useful to re-order them so as to take account of the level and type of involvement by each of the supporting stakeholders. Such a progression would start with only providing information and end up where significant incentives and/or disincentives coupled with leadership are used to make Travel Plans as effective as possible (see Table 6.2).

Table 6.2 Travel Plan support mechanisms and roles

Support mechanism	Support role	Support role type	
Provide better information.	Information provider	Encourager	Less involvement
Provide education and training for Travel Planners.	Educator	Enabler	
Encourage more involvement from public transport operators (and other potential stakeholders).	Capacity builder		
Subsidise Travel Plans.	Funder	Persuader	
Introduce regulations.	Regulator		
Modify the tax system.	Incentivisor		
Consider impact of wider policies on Travel Plan development and provide strategic direction and leadership.	Coordinator and Leader	Leader	More involvement

The following sections will therefore be as follows: information; capacity building; subsidy arrangements; regulatory mechanisms; fiscal mechanisms; institutional approaches; indirect support measures and leadership approaches.

At this point it is useful to point out that by its nature, the support measures often mirror some of the findings already reported (particularly in Chapters 4 and 5), and these instances are noted where they occur.

4 Information

Probably the simplest support role available to the Travel Plan industry is to provide information to potential (and existing) Travel Plan organisations.

Significant efforts have been devoted to developing information on Travel Plans over the years by academics, consultants, advocacy groups, charities, international NGOs, local authorities, national governments and various European Commission-funded research projects to name but a few, yet still organisations – and even some government bodies responsible for promoting them – claim ignorance of the Travel Plan concept. Clearly, then, there is a job still to be done in this area.

In particular, there is a perceived need for robust information on the costs and benefits afforded to organisations (and the wider community) by Travel Plans being implemented in various contextual conditions. Here, one issue is that standardised monitoring regimes have not yet succeeded in being adopted on a significant scale, despite some seemingly heroic attempts being made to try and make this happen – for example, MAX SUMO (see MAX Project 2009a) by the European Commission and iTrace pushed by Transport for London (iBase Systems 2008). This is likely due to issues identified throughout this book (unique circumstances at every site, insufficient resources lack of interest by organisations with Travel Plans and so on) but does mean that other organisations remain unwilling to take a risk where there is so much uncertainty as to the outcome. Specific research gaps include:

- What is the full potential of Travel Plans at the regional or national level in delivering transport and wider policy benefits?
- Under what circumstances are Travel Plans an appropriate measure to implement? Where should resources be concentrated to demonstrate best returns?
- What can we learn from examples of best practice, not only at an organisational level, but also at the local authority level? In particular, how can lessons from School Travel Plans be transferred to other Travel Planning sectors?
- What factors/pressures most effectively motivate organisations to participate in Travel Planning activities?
- What benefits from Travel Planning should organisations expect?
- Where within an organisation should responsibilities for Travel Planning rest?
- How can business travel initiatives be more effectively integrated with 'traditional' Travel Plans?

A second issue is that much of the information produced reads like it has been written by a transport planner for another transport planner to read. It is perhaps therefore not so surprising that the supposed target organisations (which usually do not appreciate the relevance of levels of congestion or air pollution in the streets) rarely read it, quite apart from act on it.

Instead, information that is produced needs to be relevant to the target audience, and deal with issues that will be in their interests to act on. It also needs to contain facts that can be used by the reader to build a robust business case within his/her organisation. Good examples do exist however (as alluded to in Chapter 5). Thus, Transport for London produces a whole series of reports on topics as diverse as fleet management, operating pool bicycles, setting up Travel Plan networks and evidence on the health impacts of Travel Plans (TfL 2011). Similarly, the National Business Travel Network and 'Best Workplaces for Commuters' provide organisation focused materials on Travel Plan and related issues (NBTN 2011; BWC 2011).

5 Capacity Building

For a relatively undeveloped sector like Travel Planning (and mobility management/TDM more generally), there is obviously a clear need to build up a means of delivering the final 'product'.

In the short-term an 'enabling' or 'facilitating' role is required, and possible services in this context could range from a public agency providing/developing information (as already discussed); to encouraging new organisations to adopt Travel Plans, establishing networks between organisations and the wider transport sector; tailoring Travel Plan instruments to site-specific requirements; developing the whole Travel Plan package for a specific site; through to implementing and/or operating Travel Plans or LTPNs on behalf of a client; and ultimately to setting out a strategic policy vision for a local, regional or national transport authority relating to Travel Plans. Fundamentally, these short-term capacity building elements are about encouraging organisations to adopt Travel Plans (the various means of which are largely covered in the other sections of this chapter); establishing support networks; and outsourcing Travel Plan tasks. In the longer term meanwhile, the challenge is to develop a core of professionals in the area to embed these short-term measures into organisation management and public policy frameworks and thus take Travel Planning on to the next level through education and training.

Because the new delivery agent being introduced into the sustainable transport sector probably has only a very limited understanding of transport (but can potentially have so much influence on the success or otherwise of a Travel Plan at a site), support with networking can be a relatively straightforward and low cost but crucial function. There are two obvious ways to do this. One is for a public authority to identify and ideally support Travel Plan champions to 'spread the word' both within organisations and in the wider world about the benefits of the concept. And a second is to provide an introductory agency to link organisations with others of similar circumstances, and or with service providers such as Travel Plan consultants, technology suppliers and particularly transport operators. Beyond this, public authorities might set up informal networks (like a Neighbourhood Transport Forum) or even more formalised networks of Travel Plan organisations across a city or in a more local area organised at Area Travel Plans or Transportation Management Associations (see Chapter 4).

Next, a major barrier to effective Travel Plans being adopted is the level of resource required – not only financial, but also in terms of personnel, time and particularly skills – and so one solution has been for local authorities and other stakeholders to outsource this and create special products and services to remove as much of the hassle of operating a Travel Plan as possible from the organisation (see Chapter 5 for more examples of this). Thus, one approach, used in Birmingham (UK), has been for the City Council to set up Birmingham Travelwise. At its most basic level this scheme encourages as many organisations to join as possible, and then provides them with a fairly generic range of services that are designed to be as simple as possible to implement. Meanwhile, another model is the Dutch approach of the so-called VCCs (Traffic Coordination Centres). These regional centres operate almost independently of local government and actively engage with local organisations to persuade them to adopt Travel Plan measures. Alternatively, public authorities have established Local Travel Plan Networks in localities to implement transport measures at a level one step removed from the organisation (see Chapter 4). Finally, in the Netherlands, the UK and particularly in the United States a number of Travel Plan service providers have emerged to undertake specific elements of Travel Planning away from organisations themselves (see Chapter 5). One example is Liftshare – a company that operates a website approach to organise lift sharing both at a regional level, but also at an organisational level (Liftshare 2011).

A further point is that because Travel Planning is a relatively new area in many countries, it is only now developing into a viable career path, even in countries (like the USA, Netherlands and UK) where it has been carried out for longer. Moreover, the area is still not fully accepted as being bona fide even by many transport professionals, the result of which is that only relatively few senior posts exist in the public and private sector meaning that many longer term Travel Planners tend to leave the sector to take up other careers. In the UK, this has been recognised as being a problem for a number of years and consequently training and education in the field is now growing. Thus, Travel Plan advocacy organisations ACT Travelwise and MODESHIFT (the national associations for Travel Planners and Travel Plan officers, and for School Travel Planners respectively) as well as some companies – such as Mayer Brown – now provide short courses on the topic. Meanwhile, the Open University provides a Level 3 NVQ Continuing Professional Development course in Travel Planning. There are also several MSc programmes that touch on relevant Travel Planning material (for example, at University of the West of England, Imperial College/University College London, Oxford Brookes University and the University of Surrey), and one that specifically specialises in Travel Plans – Loughborough University's MSc in Sustainable Transport and Travel Planning.

Elsewhere, the EPOMM Plus project is currently reviewing Travel Plan training and education programmes across the European Union and is in the process of awarding a 'quality label' to programmes that focus on developing the skills necessary for delivering mobility management in practice (EPOMM 2011).

6 Subsidy Arrangements

As is often the case in the public sector, a key barrier preventing the widespread adoption of Travel Plans is seen to be a lack of sufficient financial resources, and revenue funding in particular. This means that many authorities claim they are unable to set up new Travel Plans or even maintain links with those organisations with a Travel Plan already in place. Clearly this has severe implications for the continued operation of Travel Plans in general, and especially on their monitoring and evaluation efforts.

One approach has been to raise revenues using the planning route through impact fees or developer levies (see next section for more information on this). Thus, the so-called 'roof tax' in parts of Milton Keynes in the UK essentially replaces the typical 'site-by-site' approach to calculating planning gain with a standardised, transparent and formula-based method for calculating the fee charged. Other perhaps more appropriate 'impact fee'-style mechanisms – in that they directly incentivise developers towards adopting a Travel Plan – charge developments either for the car parking spaces provided at the site or else the number of predicted vehicle trips to be generated by the development over the course of its lifetime (Enoch and Zhang 2006). Indeed, this is already done in cities including San Francisco, California and Toronto, Ontario (Ubbels et al 2004).

A related idea in Nottingham, UK is connected to the implementation of a workplace parking levy – a local tax on organisations in the city on private non-residential parking spaces which was introduced in October 2011 and which will be used to part fund an extension to the light rail system. Here, a proposal (subsequently dropped), was that organisations operating a high quality (ultimately accredited) Travel Plan would be offered a discount on the amount of money they were liable.

Next, organisations in Scotland and London are offered consultant support in developing Travel Plans for a limited period. Such a 'Site-specific Advice' scheme also used to be available across the UK (Potter et al 2003). And, subsidy schemes allowing organisations to buy Travel Plan-related infrastructure such as bicycle racks, showers and so on for less money have been in place for several years in Nottingham and London in the UK, whilst a more recent scheme to support organisations introducing or expanding teleworking operates in Virginia, USA (see Chapter 5).

Meanwhile, additional grants can also be targeted at establishing Local Travel Plan Networks, or innovative Travel Plan related tools more generally. Such tools include joint distribution centres and delivery clubs for networks of retail outlets, car clubs, video-conferencing suites or neighbourhood 'virtual work centres'. One example where this occurs is the Allégo scheme in Montreal, Canada. Here, universities and companies are eligible for technical and financial assistance from the regional transport agency, the latter covering 50 per cent of the development, evaluation and promotion costs to a maximum of €17,000 (OECD and ITF 2010a).

In applying such subsidies, one potential way forward may be to establish Transportation Demand Management (TDM) Accredited Authorities and TDM Centres of Excellence, whereby extra money would be made available to selected or targeted local authorities to develop Travel Plans in a more comprehensive manner than currently. Too, a form of Travel Plan Bursary scheme could be introduced as a means of capacity building, as a minimum for those authorities contemplating becoming a Centre of Excellence, as could a modified Site-specific Advice (SSA) programme, whereby organisations would be offered consultant support in developing Travel Plans for a limited period.

7 Regulation

In some ways, regulation is the clearest way of ensuring the take up of Travel Plans and consequently various regulatory approaches have been tried around the world for many years. Essentially, there are two major approaches that have been adopted:

- Requirements for existing organisations to adopt a Travel Plan; and
- Regulations requiring developers of 'new' sites to introduce a Travel Plan as a condition of securing planning permission.

7.1 Regulations for Existing Organisations

As alluded to in Chapter 2, organisations with more than 300 employees (or 800 employed on freelance contracts) in Italy are required to employ a Travel Plan coordinator under the 1998 Decree on "Sustainable Mobility in Urban Areas". Known as "Decreto Ronchi", by early 2006 some 54 area mobility managers and 640 (out of 1200) company mobility managers were in post. Although there are no quantitative targets set and no penalties in place for companies not complying, there is financial and technical assistance available (OECD and ITF 2010a).

In the USA, several State/Provincial, regional and local jurisdictions formerly mandated so-called Commute Trip Reduction (CTR) programmes for certain types of employers, although many have now been abandoned. Of these, the largest and best known still operating is the Commute Trip Reduction Law in Washington State, which was passed in 1991 by the State of Washington under its Clean Air Act and introduced in 1993. Subsequently revised in 1997 and again in 2006, this law requires local governments in those urban growth areas of the state with the greatest levels of traffic congestion to develop and implement plans to reduce drive-alone trips and vehicle miles traveled per capita. Specifically, the law affects employers with 100 or more full time employees at a single worksite who begin their scheduled workday between 6 a.m. and 9 a.m., though most construction and seasonal agricultural workers are exempted, and more than 1,050 worksites and more than 530,000 commuters statewide participate in the CTR programme. In

Washington State the law requires certain planning and monitoring activities but imposes no punishment if employers fail to meet the planning objectives.

Kadesh and Roach (1997) reports that as of 1995 every affected city and county in Washington had passed a CTR ordinance, and that 100 per cent of the 825 affected worksites in the state participated in the programme. Moreover, more than 30 private and public employers joined voluntarily, as did 133 smaller state agency worksites not subject to the legislation, bringing the number of employees covered by CTR as of December 1995 to a total of 465,000 (about 20 per cent of all employees in the affected counties).

In terms of results, in the first two years WSEO (1995) finds overall single occupancy vehicle use fell by 4 per cent and vehicle miles travelled by 6 per cent, and that 36 per cent of worksites met their 1995 SOV or VKT goals. Meanwhile, an estimated 128 million vehicle kilometres per year during peak commute times were eliminated amounting to reduced annual emissions of carbon dioxide of 29,900 tonnes. More recently, WSDOT (2011) states that across Washington State there were nearly 28,000 fewer vehicle trips each weekday morning in 2009 made by employees commuting to CTR worksites, resulting in a reduction in travel delay of around 7.5 per cent in the Central Puget Sound Region saving each morning peak traveler $USD59 (€40) a year in wasted fuel and lost time. Meanwhile 62 million vehicle miles travelled were cut each year between 2007 and 2009, equating to 27,490 tonnes of carbon dioxide per annum.

Rye (1999) finds several reasons for the survival of the legislation which include the liberal politics of the state; 'locally-owned' nature of the law (the CTR law originated at state rather than federal level); and the triple emphasis in the CTR law on reducing traffic congestion and energy consumption as well as air pollution, which has 'broadened its appeal'. In addition, Rye suggests that high economic growth in the region coupled with a growing transport problem helped persuade employers of the benefits of the law to them.

Meanwhile, the Department of Transportation in Delaware requires and enforces traffic mitigation agreements that include TDM requirements and that the programme "has teeth" (NCHRP 2010: 4), which is unlike many other similar regulatory environments.

7.2 Regulation through the Planning System

Travel Plan regulation through the planning system is perhaps best developed in the UK, though there are a number of other examples that can be thought of as being 'good practice'. Thus, MAX Project (2009a) cites several cities in this regard including Zurich, Switzerland, Portland, Oregon and Dublin, Ireland.

Looking to the UK, organisations needing planning permission to expand their existing site or develop a new one can be required to establish a Travel Plan. In England and Wales, national guidance on how this should be applied has been laid down in recent *Planning policy guidance notes* 13 (PPG 13) – on transport and land use (DoE and DoT 1994; DTLR 2001). This guidance has attached significant importance to Travel Plans and advises that local authorities should include Travel Plans as a part of their own planning policy, and should also require all large scale employment, retail and leisure developments to have a Travel Plan (Rye 2002).

However, the fact that regulation is limited only to development decisions means that more often than not it is suburban and city-edge sites that are required to implement Travel Plans, not necessarily the companies in areas which would benefit the most from the introduction of the measure. In addition, there are several other problems as far as using planning agreements to limit traffic growth. One is that the planning rules are implemented purely at discretion of local authorities, and although PPG 13 is now increasingly encouraging more standardisation in how these are applied, there are still issues. Thus:

> As with many other [transport] demand management measures, competition between adjacent localities for economic strength can seriously reduce the effectiveness of well-intended policies. Given the choice between accommodating the requirements of a major project, which will enhance the local economy, by relaxing their more stringent policies or maintaining those policies and seeing the project go elsewhere, many authorities will opt for the former. While a firm national, or regional, policy framework might help to avoid such 'bidding' situations, it would be at some cost to local autonomy on key decisions (IHT 1997: 293).

At the practical/detail level too, there are difficulties. For example, there is relatively little data on how effective Travel Plans are – making it difficult for local authorities to judge whether a Travel Plan is good enough or not. In addition, developers do not often know how to write Travel Plans, leading to negotiations and delay while the plan is improved. Finally, many development proposals are speculative and transport is dealt with at the outline planning stage. This means that often there is not a very high level of detail about the type of development and the number of employees, which adds to the difficulty of drawing up the Travel Plan (Rye 2002).

Recent research referred to during interviews for Enoch and Ison (2011), suggests progress in how local authorities have used the planning system to encourage the development of Travel Plans has been disappointing. In particular, "the quality of monitoring information is poor even with those authorities that are furthest ahead", a situation the interviewee suggests was down to "resources". Specifically, "what has become very clear is that unless [the monitoring of Travel Plans] is part of some assessment process; it will go on the back burner … because local authorities deal with what is on their radar [and allocate] resources for assessment processes"[1] accordingly. This is corroborated by Rye et al (2011). This finds that local authorities need to devise a workable enforcement plan for organisations that breach their planning obligations or agreements, a task that is

1 No page number available yet as paper is in press.

especially urgent given that only four respondents of 86 had ever taken enforcement action while the vast majority had no idea how they would go about doing so. This was despite 46 of the 86 respondents stating that there were examples in their areas of Travel Plans that had not been implemented or were likely to be in breach of their condition/obligation for some reason. Finally, Hendricks (2008) draws on US experience to set out four challenges to incorporating Travel Plans land development process, namely:

- Poorly worded laws allowing an oversimplification that Travel Plans can be equated with reducing travel.
- Travel Plan processes are considered too late in the land development process.
- Balancing mobility with accessibility as interest in public transport-oriented development and pedestrian-friendly communities grows – pitting the interests of state and local government against each other.
- Current law and traffic analysis methods and standards prioritise transport levels of service experienced by motorists over that of alternative mode users.

Fundamentally, though, despite the problems outlined the introduction of Travel Plans through the planning system is likely to (and should) remain a key instrument for local authorities for a number of reasons.

Firstly, planning decisions are intrinsically linked to the transport performance of a particular area, meaning that the earlier a Travel Planner can become involved in the planning process the more influence s/he will have. In practice, in the UK it is widely recognised that Travel Plans are not considered early enough in the process although other countries have been more successful (the Netherlands in particular).

Secondly, the planning process offers significant flexibility in how Travel Plan requirements may be implemented. Thus, in the mid-2000s Transport for London was considering a system whereby every development must consider Travel Plan issues before planning permission is granted by the planning authority although it was not eventually implemented in that form. Under this proposal a single page check list demonstrating that travel impacts had been thought about was required for the smallest schemes, and a standardised Travel Plan was available for medium-sized projects, while the largest developments still needed to submit a full site-specific Travel Plan. One development of this could be to use planning gain arrangements from particularly large schemes to develop the core of a Local Travel Plan Network for the local neighbourhood. This might function rather like the 'anchor store' concept in a shopping centre, whereby many small shops benefit from the high trip attracting nature of one or more large department stores located within the development.

Thirdly, the planning decision stage is currently the only point at which a local authority can influence how a development will operate over its lifetime.

The other side of the coin meanwhile, is to ensure that Travel Plans introduced via the planning (or any additional regulatory) system include a budget to cover the costs of monitoring activities by the local planning authority.

7.3 Perspectives on the Future of Regulation as a Support Mechanism

Looking to the future on regulation more generally, Dill (1998) identifies ten issues that influenced the history and demise of mandatory Travel Plans in the USA (see Table 6.3). These include employers not accepting issues like air pollution as their responsibility, unrealistic goal setting, inflexible, unfair and burdensome regulations and relatively unimpressive results.

Table 6.3 Problems with the regulatory approach

Issue	Description
Problem definition	Prior to mandates, employers responded to externalities – that is because of self-interest. Mandates forced employers to respond to problems like air pollution that were not so obviously 'their problem'.
Goal setting	Often perceived as being unrealistic.
Adoption and implementation process	Employer 'buy in' is crucial to success, yet often policy makers ignore this.
Regulatory approach	Strict approach based on experiences with stationary pollution sources and inappropriate for transport. Seen as being too one-sided with employers taking most of the burden. Differing contextual circumstances amongst employers indicated far more flexibility needed. Regulations led to employers trying to satisfy regulations rather than implement the most effective measures.
Programme results	Regional reductions in car trips of 2 to 5 per cent commonly cited but not seen as sufficient by business sector complaining about high costs and bureaucratic hassle.
Costs	Unclear methodologies confused issues. Unit measurement of cost per employee dominant, but additional results on congestion or air quality benefits less often cited (more difficult and time consuming to generate).
Private sector role	Employers unsure as to their appropriate role in influencing employee commute habits. Differences in interests between developers and employers. Employers seen to have more influence on employees than government.
Other players	Mixed response to Travel Plans from labour unions. Little support from wider Travel Plan industry (consultants and service providers).
Policy fragmentation	Often mandatory programmes not well integrated with other potentially complementary policies and programmes.
Context	Difficult to impose during times of economic recession.

Source: Dill (1998)

Looking to alternative approaches, Rye (1999) compared experiences of using regulatory schemes with voluntary and fiscal schemes in the USA and the Netherlands (see Table 6.4). His paper agrees with Dill on the problems of regulation and in the end favours a more fiscally-oriented approach.

Table 6.4 Voluntary, regulatory and fiscal regimes compared

	Advantages	Disadvantages
Voluntary	Motivated participants; No need to regulate; No cost to public purse; Benefits for employers – to image and solves site-specific problems.	Low take up in most areas; So low overall effectiveness; No consistent modal split targets; Few economies of scale in providing services to groups of employers.
Regulatory	Affects all employers in an area who meet all pre-determined criteria; Consistent modal split targets; Can cover large areas; Economies of scale in providing services to groups of employers.	Administrative costs; Perceived burden on and so resentment from employers; Difficulty of regulating lots of employers without large bureaucracy; Problem of how to regulate small employers.
Fiscal	Can incentivise employers to adopt ETPs; Can target both employers and employees directly; Potentially tradable; Can be added to existing tax system; Can cover employers of all sizes.	May increase overall tax burden; Employers may seek ways of avoiding payment; May cause resentment amongst employers; No consistent modal split targets.

Source: Rye (1999)

Finally, Pagano and Verdin (1997) reports in a study conducted in Chicago that eliminating government mandates reduced the cost of implementing an employee trip reduction programme by half. It recommends establishing tax incentives supported by an aggressive education campaign instead.

That said, there have been moves by local and provincial government associations in the Netherlands towards including the environmental impacts of business travel and commuting journeys within the purview of the Environmental Law – which sets limits on the amount of environmental damage that organisations can inflict on society. So far a national scheme whereby organisations are awarded points for reducing their environmental impacts (including those from transport) has been resisted by business groups, but the City of Amsterdam is one municipality that has implemented such a regulation (Sonderse et al 2007). Instead, at the national level a 'covenant approach' has emerged whereby the top 50 largest companies agreed to voluntarily adopt Travel Plans with the aim of reducing single occupancy car use by 5 per cent should the regulations be dropped. So far, however, single occupancy car use has fallen 1.5 per cent – a figure which is not large enough to satisfy the Dutch Government but is too large to ignore and so the implementation date for the regulations has been postponed until 2013 (Metz, F. 2011. Personal communication, Director, KPVV, Amsterdam, 24 February).

Such an approach may be possible elsewhere, although it should be noted that in continental Europe the employer is 'responsible' for the employee's commute trips, which is not usually the case in English-speaking countries for example. Moreover, 'in all situations regulation' tends to generate a view that Travel Plans are a cost – and not a benefit to be exploited – providing a contradictory message to the information and exhortation approach and leading to a counterproductive attitude. Nevertheless, if climate change and the security of energy supplies continue to be seen as reasons for regulation at the European level, the pressure for government and local authorities may well one day result in some form of organisational and/or personal carbon allowance, and at that point travel would certainly need to be incorporated into any such scheme.

As a minimum, there would seem a strong case for allowing local authorities the power not only to retain the planning system as a means for implementing Travel Plans but also for extending them further – especially in terms of allowing Travel Planners to intervene earlier in the planning process.

8 Fiscal Measures

Given the issues with the regulatory approaches, an alternative potentially 'hard-hitting' form of direct intervention is the use of fiscal measures, either through tax incentives or tax disincentives.

8.1 Personal Tax Regimes

Van den Branden et al (2000) reviews personal tax regimes (the tax type most directly linked to commuting and to Travel Plans). This reports that countries including the UK, Austria, Greece, Ireland, Italy, Portugal, Spain and the USA view commuting as a personal expense, while others such as Belgium, Denmark, Finland, France, Germany, Luxembourg and the Netherlands, view commuting as a tax deductible expense. It adds that most taxation regimes (of both types) provide contradictory signals regarding 'greener' forms of transport, but that some countries have started to use elements of their personal taxation system to encourage the use of more sustainable transport modes. These include the USA, Ireland, the Netherlands and the UK.

In the USA the idea of targeting tax concessions specifically for company TDM measures reflects the belief that the use of direct financial incentives forms a key element in reducing single car occupancy travel to work. This concept first emerged during the mid-1980s and has steadily evolved over time. Currently, the measures include a range of options whereby employees benefit financially not just from tax relief on parking (as in the UK, for example where parking is not taxed despite being a benefit), but also from using alternatives to the car (see Table 6.5 for a summary of these).

Table 6.5 US tax position summary for employer TDM measures[2]

Option	Employer tax benefit	Employee tax benefit
Employer paid: employers provide employees with up to $USD105/month (€71)/month to commute via public transport or vanpools (typically through passes or vouchers) and/or up to $USD200 (€135) a month to pay for parking.	Compared to providing a salary increase, the employer saves on payroll taxes. No payroll taxes are paid on the value of the benefit.	Employee receives up to $105/month (€71) tax-free to commute via public transport or vanpool (and $200/month (€135) tax-free for qualified parking). The employee does not pay any taxes on the value of the benefit.
Employee pre-tax deduction: employers allow staff to reserve up to $USD105/month (€71) in pre-tax income to pay for public transport or vanpools and/or up to $USD200 (€135) to pay for parking.	Employer saves on payroll taxes. No payroll taxes are paid on the income that is reserved by the employee.	Employee saves on income tax and payroll taxes. The amount of the benefits is no longer treated as taxable salary.
Employee, employer share costs: employers pay public transport, vanpool or qualified parking costs up to a certain limit and the employee pays for the remainder of the costs through a pre-tax deduction. Total employer and employee paid costs cannot exceed the monthly limits of $USD105/month (€71) for public transport and vanpool and $USD200 (€135) for qualified parking.	Employer saves on payroll taxes on the portion of the subsidy paid by the employee through a pre-tax deduction, and on the value of the benefit provided to the employee compared to providing a salary increase.	Employee does not have to pay any taxes on the portion of the benefit provided by the employer. Additionally, the employee saves on income tax and payroll taxes by taking a pre-tax deduction to pay for the remainder of his/her commuting costs.

Source: BEST Workplaces for Commuters (2005)

As a result of these benefits being established a new Travel Plan services industry has developed, one example of which is the *TransitChek* scheme used in New York and elsewhere in the USA. Here, more than 14,000 employers purchase Metrocard passes and/or vouchers from the New York TransitCenter, an alliance of public transport operators that administers the programme for the

2 Note: The employer can also deduct the cost of the benefit as a business expense for purposes of calculating corporate income taxes, in the same manner that salary payments to employees can be deducted.

region which was first established in 1986. These cards and/or vouchers are then distributed to employees. Interestingly, the scheme is marketed as a staff benefits package rather than a transport programme (TransitCenter 2005). In evaluating the scheme, Litman (1997) noted that around 25 per cent of the TransitChek recipients previously commuted by car alone. A small amount (about 4 per cent) were previously car passengers and about 2 per cent previously walked. In the USA context he saw this as a good result. He suggested that a higher uptake by car drivers could be expected in places with a better public transport infrastructure and where companies linked TransitCheks to Travel Plans.

Ireland has subsequently followed a similar approach in a process beginning in 1999 (Taxsaver.ie 2010), while in the UK the 2001 and 2002 Budget reports moved fiscal policy from being biased towards the car instead of 'greener' modes, in the direction of a more 'mode neutral' position. Finally, the Netherlands has slowly shifted from providing a blanket concession that mainly favoured car users towards a system where public transport use is incentivised (Potter et al 2006).

In summary, Potter et al (2006) states that:

> General commuting trip tax concessions tend to stimulate car commuting and trip lengthening, and are costly to the state concerned. Similarly for general public transport commuting trip tax concessions
>
> In countries where commuting costs are not tax-deductible, targeted tax concessions upon employer-provided Travel Plan benefits have helped support the development of effective Travel Plans and hence influenced modal shift.

In addition, the paper notes that tax concessions for employers are needed as well as those for employees, and that the corporate taxation system in particular has much untapped potential.

8.2 Corporate Tax Approaches

Accordingly, more recent work has developed a series of possible tax changes in other areas including the corporate tax regime, specifically business rates, capital/industrial business allowances and Travel Plan tax credits (Potter, S. 2010. Personal communication, Professor of Transport Strategy, Open University, Milton Keynes, 6 January).

- On business rates (a business tax based on the value of the property used), he notes that organisations could be taxed according to the number of parking spaces they had in place – although there could be significant boundary effects as rates would probably need to be differentiated by location. In addition, there could be scope for similar but perhaps more radical discounts with the proposed introduction of a Supplementary Business Rate.

- Second, businesses could be offered tax relief on the amount of money they spent on Travel Plan measures, for example bicycle parking, bus shelters, cycleways, car park charging equipment or car sharing bays. Here, though, there is a problem that Travel Plans are generally not particularly capital intensive – instead they tend to need revenue to be spent and so any change would have only limited impact.
- Third, tax credits could be offered relating to the non-capital expenditure of a Travel Plan, whereby companies could, for example deduct 125 per cent or 150 per cent of its expenditure on a Travel Plan when it determines its profits, as opposed to 100 per cent normally. Here, though, the measure would be output not outcome based, and the spend would need to be additionally audited. One other related approach could rely on a system of accredited Travel Plans whereby discounts of say 5 per cent, 10 per cent and 20 per cent could be given to organisations in receipt of a bronze, silver or gold rated Travel Plan. This would reward outcomes and would be audited by a separate organisation.

Overall, corporate tax benefits would only apply to private enterprise and not to the public sector. Moreover, many companies already take full advantage of any available allowances.

Finally, beyond these corporate tax approaches, there are several more radical options that might also be considered (*op cit*). These could include providing:

- Tax benefits for employees living near to their workplaces.
- Tax benefits for employers providing accommodation for their employees (within a limited distance of the work site).
- Tax breaks for communication equipment such as tele-conferencing and video-conferencing suites, neighbourhood work centres and so on, designed to minimise the need for people to commute and make business trips.

8.3 Determining Fiscal Policy

Essentially, in the UK at least, most of the fiscal measures introduced so far by National Government have been designed to maximise behavioural change while minimising 'deadweight' (that is, rewarding people for continuing their existing behaviour). A side effect of such measures however has been to make the regulations extremely complicated and difficult to understand and thus 'take up' (and the resulting impact on travel behaviour) has probably been less than hoped. Worse, for a long while there was little information available from either the Department for Transport or Her Majesty's Revenue and Customs of how organisations can benefit from such tax measures, or on how the measures actually work, and the information that is available was difficult to find.

Overall, then, it is clear that tax changes can influence travel behaviour in organisations, but that the key decision facing any government is whether it is prepared to risk an increase in 'deadweight costs' in order to generate a potentially significant benefits in terms of cutting car use. One suspects that in the current economic climate the answer will be a clear 'no'. Indeed, recent indications from the UK Treasury suggest that some of the green commuting tax benefits offered on cyclist breakfasts, cycle to work and bus season ticketing schemes are to be cut in a bid to simplify the tax system and save public money.

Interestingly, the final point made in Potter et al (2006) is that minor changes to the tax system in themselves will not do much to address transport and environmental objectives, unless these objectives are in line with wider government policy moves towards the same goals. However before these are considered it is first necessary to examine the institutional issues.

9 Institutional Measures

Given the centrality of Travel Plans being a delivery mechanism, it makes sense to consider how they themselves are delivered, although in practice this aspect seems to have been rather neglected.

Looking first at the functional or 'horizontal' dimension, unsurprisingly in most local authorities responsibility for Travel Plans rests with transport and/or planning teams, rather than with environmental or business units, for example (as in many organisations – see Chapter 5) which may well make some organisational sense but which also indicates a potential loss of opportunity.

In the vertical dimension meanwhile, as already noted in Chapters 2 and 4, Travel Plans (and Local Travel Plan Networks or Groups) provide an additional level of delivery for transport policy, which Enoch et al (2005) describes as being the 'tactical' level (see Figure 6.1).

From this model, it is helpful for regional and local authorities to see themselves as being 'strategic' and the LTPN/Travel Plan organisations as being 'tactical' delivery agents where the strategic bodies are responsible for applying wider strategic policy objectives and implementation frameworks to the local context and distributing the allocated resources, while the tactical bodies actually implement the Travel Plan measures on the ground. Meanwhile the point where the two levels interact can be termed the 'interface'.

Enoch et al (2005) goes on to propose that two models are available to local authorities with operational LTPNs: a 'federalised' structure or a 'centralised' structure. The characteristics of both are presented in Table 6.6.

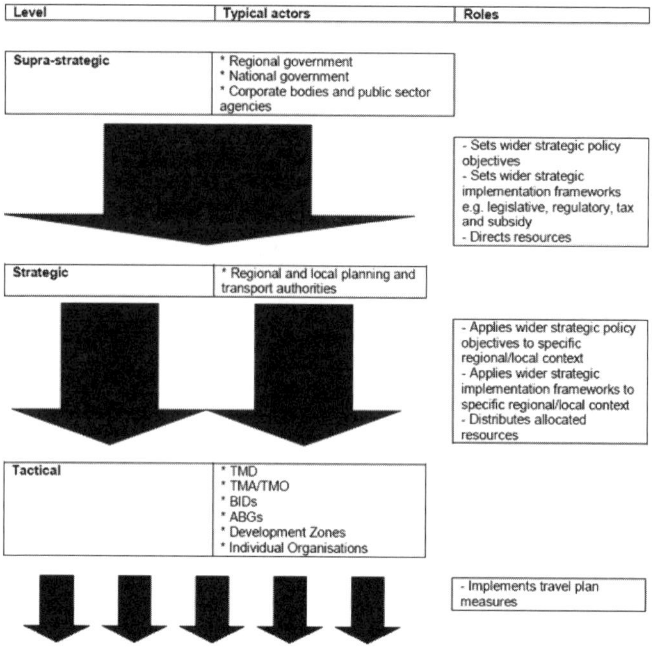

Figure 6.1 Travel Plans as the 'tactical' level of delivery

Table 6.6 Federalised and centralised interface model attributes

	Federalised	Centralised
Strategic body	As a facilitator; Weak; Sets up a network of tactical organisations; May provide financial incentives; Provides advice and encouragement; Facilitates contacts; Facilitates meetings.	As a leader; Strong; Establishes group for tactical organisations; Provides financial assistance; Provides advice and encouragement but may also use regulation as a stick; Establishes contacts and negotiates deals; Leads meetings.
Tactical organisation	As a member; Strong; Joins network; Raises its own revenue; Implements measures if in its own interest; Makes its own contacts; Contributes to meetings.	As a 'customer'; Weak; Joins group; Draws on club resources; Buys services from strategic body if in its own interest or to meet regulatory targets; Makes use of existing contacts and negotiated deals; Attends meetings.
General comments	Process is bottom-up; Relatively small number of active/enthusiastic members; Large per member impacts.	Process is top-down; Relatively large but mainly passive/reactive membership; Relatively small per member impacts.

In reality, Enoch et al (2005) notes that the federalised interface structure tends to be applied in situations where there are fairly small numbers of large, powerful, influential and cohesive local groups/organisations, while a centralised structure usually applies where the opposite conditions exist. It then presents examples of the federalised model being in operation in Nottingham and Bristol (UK), in Boulder, San Diego and Montgomery County (USA), and in Toronto (Canada); and writes about Birmingham's Travelwise (UK) and Rotterdam's VCC (Netherlands) programmes as centralised model cases. Further illustrating this idea, NCHRP (2010) reports that New Jersey uses TMAs across the whole state to deliver TDM services while in London, Tyler et al (2007) points out that there are examples of four of the five LTPN 'types' extant with a high degree of flexibility being available to local transport officers in selecting which may be the most appropriate structure to use.

Meanwhile, a second important element that emerges from Enoch et al (2007), is the level of involvement of the planning or transport authority at the strategic level *vis a vis* the tactical organisations, which ranges from a highly interventionist to a more laissez faire approach. So, in the UK context Cairns et al (2005) suggests that York City Council is an exponent of the former approach and effectively chooses to work intensively with a relatively small group of organisations, while Birmingham City Council is far less 'hands-on' and encourages companies to use its centrally administered Travel Plan services as and when required (Cooper, M. 2002. Personal communication, Project Leader, Birmingham Travelwise, Birmingham City Council, Birmingham, 20 November).

More broadly, institutional measures also relate to the leadership measures (see later) regarding the role of (and vision for) Travel Plans. Acknowledging this, one key proposal that pulls together many of the above ideas put forward in this section (and indeed in the chapter so far) is to set up a number of TDM Centres of Excellence. Such a model, whereby extra resources are granted to local authorities to implement mobility management measures (and Travel Plans in particular), into the local consciousness and thus reduce reliance on the car by implementing policies tailored to meeting the needs of local neighbourhoods initially in the areas of greatest need, but ultimately across the whole of the local authority area. Crucially, these Centres would be able to attract additional sources of funding and perhaps be able to develop new financing streams if necessary. There may also be scope for such Centres to be granted further regulatory powers (see earlier), which would most be granted through the planning system, although further environmental or car parking powers may be appropriate in some cases. Such Centres could also be given a regional role in helping neighbouring local authorities that are 'sympathetic' to mobility management introduce and operate their own Travel Plan strategies, which could subsequently aim to become TDM Accredited Authorities, and ultimately possibly even Centres of Excellence in their own right. Such accreditation could be awarded either by government or professional bodies or more likely by a Travel Plan advocacy organisation.

Alternatively, there may be scope for tying such an award to an already existing environmental or corporate social responsibility labelling system.

10 Indirect Support Measures

Such support measures as already described being in place would likely see Travel Plans be more effective, but as several interviewees suggest in Enoch and Ison (2010), there is also a real need for even wider supporting measures to 'lock in' the benefits afforded by these interventions. This section therefore considers what these might be.

10.1 Information and Awareness Campaigns

Information measures here suggest the need for wider government campaigns educating the public as to the threat of global warming and the benefits of using alternatives to the car as, for example in the UK Government's *Act on* CO_2 campaign (DirectGov 2011). More focused measures may include highlighting the costs that businesses pay for parking as a proportion of a business tax for example, or else applying the type of Individualised Marketing campaigns used in London, Japan and Australia on a broader basis.

10.2 Subsidy Schemes

Subsidy-led approaches supportive of Travel Planning include direct subsidies for Travel Plan activities, as well as broader investments enhancing non-car modes, such as in walking or cycling infrastructure for example. Similar investments made in supporting bus services tend also to help the effectiveness of Travel Plans. Thus, the January 2008 announcement by the UK Government that an additional £GBP110 million (€122 million) was to be invested to create a further ten Cycle Demonstration Towns and one Cycle Demonstration City, on top of the existing six demonstration towns, would be a good example of this (DfT 2008).

10.3 Regulatory Options

Regarding regulation there are three obvious methods for supporting Travel Plans – through planning, environmental and parking policy. Of these, the former two options have already largely been discussed, but there is significant scope in the latter.

It is no coincidence that Travel Plans and parking are closely related. In essence, too little parking supply for the level of demand requires organisations either to provide more supply or else to manage the parking they have and/or provide alternative solutions (a Travel Plan), while too much parking supply for the level of demand represents a direct cost to the provider and so should be a rare occurrence in an economically rational world. The 'right' level of parking on the other hand represents a situation whereby the access needs of the organisation are (apparently) satisfied by the existing transport system. From this, Travel Plans tend to work best when the provision of parking is tightly controlled. One direct way of achieving this could be to impose a 'parking cap' on a particular area, development or organisation as occurs in Boston and Portland in the United States (see Enoch et al 2006). Another approach is used in Santa Monica, California, where the local council has introduced a mandatory Parking Cash Out programme (United States Environmental Protection Agency 2000). This applies to firms employing 50 or more people that have at least one work site located in the South Coast Air Basin who:

- do not own their own parking (normally this means parking is leased);
- subsidise employee parking (that is, pay all or part of the cost of parking for any employee);
- can calculate the out-of-pocket amount paid for subsidised employee parking (for example parking costs are unbundled, or separated in the lease agreement); and
- can reduce the number of parking spaces leased without penalty (by not having to break the lease or pay for unused spaces for example).

Such employers are required to offer all eligible employees the option – in lieu of subsidised parking – of a cash allowance equivalent to the parking subsidy. Eligible employees are those to whom a subsidised parking space is available, even if it is not used. The parking subsidy is equal to the out-of-pocket amount paid by the employer for employee parking minus any contribution by the employee and minus any commute-related subsidies otherwise given to the employee (for example, for public transport or lift sharing). Employees have the choice of foregoing the cash offer and continuing to receive subsidised parking or accepting the offer and either paying for parking themselves or finding an alternative way to commute to work. Santa Monica recognises parking cash out schemes for their contribution to overall trip reduction and to the emissions reduction requirements of the area's Clean Air laws. Parking cash out schemes have also been introduced in the UK, though the practice is not widespread (Enoch 2002).

One other parking-based approach is to require all new developments to provide parking spaces as currently, but then transfer the control of these spaces to the local authority thereby effectively abolishing the creation of any new private non-residential spaces.

Slightly different is to increase support for public transport services. This can be done by using municipal regulations to require employers to refund a proportion of the cost of public transport season tickets to their employees as in Vienna, in cities across France and in Brazil (Ubbels et al 2004). Or, local authorities could include Travel Plan related elements when tendering for an operator to run specific public transport service franchises.

10.4 Fiscal Measures

Regarding fiscal measures, clearly the introduction of road user charging either nationally as proposed from time to time in the Netherlands, or locally in more towns and cities would strongly reinforce the role of the Travel Plan. Once again there are also possibilities in targeting parking provision. Thus, VAT (which in the UK is currently not applied to parking spaces) could be re-applied. Alternatively, an option could be to re-balance the Business Rates to increase the value placed on parking spaces while possibly reducing that placed on less environmentally damaging aspects of a building. Finally, parking spaces could be taxed, either through the revenue of commercial parking operators as in the United States, or through employers as proposed with the workplace parking levy (WPPL) in Nottingham in the UK (Enoch and Ison 2006). Interestingly, as noted earlier, the link between the WPPL and Travel Plans was actually made in Nottingham through proposals which suggested that organisations with Travel Plans in place be made exempt from paying it, although this has not been adopted in the latest documentation.

10.5 Institutional Measures

Continuing with the parking theme, an institutional measure is to link Special Parking Areas – that is areas where parking offences have been decriminalised and are now enforced by the local authority – more closely with Travel Plans. This allows revenues gathered to be earmarked directly into Travel Plan improvements therefore reinforcing the acceptability of both the parking regime and Travel Plan and is already used in parts of the United States where they are known as Parking Benefit Districts (see Shoup 2005).

Another approach, meanwhile, is to see Travel Plans as a means of supporting other transport policies. In this regard, the UK Highways Agency has done just this through its Influencing Travel Behaviour programme, whereby large developments located near to the highway network may be required to introduce a Travel Plan in return for securing the consent of the Agency in the planning permission process (HA 2011). In this way, the idea is that the need for increasing road capacity can be avoided for as long as possible.

Other wider institutional measures could include:

- modifying project appraisal criteria for transport planning, highway and railway schemes to include the potential role of Travel Plan (and wider TDM) measures where appropriate;
- explicitly including Travel Plan elements such as the cost of car parking and business travel in auditing procedures for local authorities, government departments and public sector bodies; and
- considering how Travel Planning measures might be incorporated either into a future development of the European Union Greenhouse Gas Emission

Trading Scheme (at a company level) or else in any personal carbon trading scheme arrangement.

- increasing the profile of Travel Plans within the transport planning process, such that they could become a primary delivery mechanism of transport improvement measures at the neighbourhood level;
- as part of this, there is a need for the performance of Travel Plans to be benchmarked against other transport policy mechanisms, ideally by using the same (albeit suitably modified) assessment criteria as used to judge more traditional transport planning, highway and/or railway infrastructure schemes.

10.6 The 'Do-nothing' Approach

Ironically, one of the most effective wider support measures for government to take may well be to do nothing but let the transport system continue to become more congested and less reliable. Anecdotal evidence would seem to indicate that there is a steady shift occurring in how people perceive driving, from it being something they enjoy to something they have no choice in doing. For that reason, if for no other, Travel Plans somehow need to be able to capitalise on this feeling by offering people a real alternative to them always taking their car.

Unfortunately, however, while cynically speaking 'do-nothing' may not be a disaster for Travel Planning, there are many instances where government decisions in related or even other apparently unrelated sectors actually do make things dramatically worse. Two UK examples of the former experience have concerned parking. First, in the review process for the revising Planning Policy Statements the UK Government began to abandon the concept of maximum parking standards in some instances (Local Transport Today 2008), while a Government announcement made during Spring 2010 stated that hospitals were no longer to charge for car parking (although financial circumstances subsequently persuaded the newly elected Conservative Liberal Coalition Government to reverse this) (BBC 2011). Meanwhile the (theoretically unrelated) policy of allowing parents to choose which school to send their children to can also have significant implications for car use and carbon dioxide emissions (Marshall et al 2010; Van Ristell, 2011), as can changes in how healthcare is delivered, and by extension for Travel Plans.

Instead, governments need to recognise that as a minimum Travel Plans are complementary to supply-side solutions and where providing more road, rail or air capacity becomes ever more difficult due to the changing political, economic, social and environmental climate, they offer an alternative (and more acceptable) way forward.

11 Leadership Measures

That such injurious policy decisions can be taken serves to underline the importance of the final support measure, which once again is highlighted in Enoch and Ison (2010), specifically the lack of leadership and strategic direction offered by national and local government. Such a function has long been recognised as being crucial for transport as a whole (though often apparently ignored in reality in some countries). Arguably, however, it is even more important for 'new' measures such as Travel Plans, particularly given the strong 'delivery agent' and 'package of measures' themes coupled with the fact that the stakeholders are usually 'outsiders'. It is also clearly necessary that measures such as those already discussed in this chapter will be much more effective if they are introduced in as timely and coordinated a fashion as can be managed.

The importance of leadership in promoting and embedding Travel Plans has been recognised from time to time, though not yet on a comprehensive basis. Thus, as noted in Chapter 3 the UK Government used its 1998 transport White Paper to require its own buildings to set up Travel Plans (DETR 1998), and for similar reasons local authorities, the NHS, universities and some leading private sector organisations in the UK have adopted Travel Plans. Meanwhile Travel Plan lobby groups around the world such as the Association for Commuter Transport in the USA and Canada; KPVV in the Netherlands and ACT Travelwise, the National Business Travel Network and the Campaign for Better Transport in the UK have been trying to persuade Government ministers to do more in this area. Specifically, one suggested approach has been to try and get UK Government ministers to make speeches extolling the benefits of Travel Plans, and slightly more targeted, to have them meet with business leaders to persuade them of the benefits of adopting Travel Plans – a process that is slowly beginning to happen in the UK context.

Beyond these cases, though, no obvious examples present themselves in this area. That said, there are some ideas worth considering in this regard.

One way would be for senior government ministers to raise the idea of Travel Planning with heads of organisations, encouraging them to directly raise the profile of Travel Planning within their companies. This could be 'sold' as being a company benefit not only in terms of direct cost savings, but also in wider public relations terms. Such a proposal would be most effective if linked into the role of existing Travel Plan advocacy organisations, which has already opened communications between business and government and also if coupled with measures to support capacity building initiatives within the sector.

At the same time, there could be scope to highlight the benefits of Travel Plans more generally among the business community – including those which stand to benefit most directly, namely public transport operators and other transport suppliers who really must be convinced that it is in their own interests to develop new products to cater for the Travel Plan market (as described in Chapter 5).

Such developments would be all the more effective if they could be tied in with the launch of a package of support measures (for example subsidy programmes and tax breaks and perhaps even a re-launch of the Travel Plan 'brand'. Such a re-launch would be particularly effective if it included some form of standard or accreditation for Travel Plans (see earlier).

Finally, as already noted the importance of identifying policy champions to drive Travel Planning forward needs to be recognised and acted upon. It is suggested that such people can be found at different levels – right from the part-time Travel Planner in a small organisation; up through local authority Travel Plan officers and Travel Plan consultants; to the major decision makers in multi-national boardrooms and in government.

12 Summary and Conclusions

As implied throughout this chapter, interventions supporting Travel Plans can clearly make a significant difference in determining the success of Travel Plans in practice. However, arguably just as crucial (if not more so) is the role of indirect support measures.

Indirect support measures refer not just to wider transport policy measures but also social, environmental and especially economic policies (and indeed any other societal factors although these are not always under the control of policy makers) that potentially impact on the travel behaviour of individuals. This is because often the impacts of such factors can be degrees of magnitude greater than the direct interventions discussed previously. Thus, for example, the Ash Cloud that arose from the Eyjafjallajökull Volcano eruption in Iceland in April 2010 grounded 108,000 flights, disrupted 10.5 million passengers and cost the airline industry at least $USD1.7 billion (€1.1 billion) in lost revenue, (Budd et al 2011), while the Global Economic Recession has slowed traffic growth across the world (OECD and ITF 2010b). Meanwhile analysis of the UK School Census by Loughborough University PhD student Jessica van Ristell reveals that the policy of allowing parents to send their children to the school of their choice results in a reduction in total traffic on school days of around nearly 1 per cent – an effect that is possibly more pronounced in generating car use than school Travel Plans are at reducing it (Van Ristell 2011)!

Within transport policy, too, there are notable examples of where policy decisions would likely affect levels of car use (and thus of Travel Plan effectiveness), essentially because if people are 'persuaded' in some way not to use their cars then they are more likely to adopt any measures put in place by a Travel Plan. Thus, for example, in the last years of the twentieth century successive UK Governments applied a 'fuel tax escalator', whereby tax on petrol and diesel was increased by 5 per cent (later 6 per cent) above the rate of inflation in order to restrain car use and raise money for the Treasury, a policy which Glaister (2002) credits as halting the growth in car use in 1999 and 2000. However, when fuel prices rose dramatically during the summer of 2000 then protests by farmers and lorry drivers in September of that year forced the UK Government to abandon this approach, which in the long-run will mean traffic will grow by nearly 2 per cent and fuel consumption 4 per cent more than it would otherwise have done as a consequence (*op cit*). Similarly, road pricing schemes as adopted in London, UK, Stockholm, Sweden and Valletta, Malta can be expected to make Travel Plans more attractive to employers and their staff.

Table 6.7 pulls together the possible types of direct intervention available to the Travel Plan industry and arranges them firstly as to whether they act as incentives (carrots) that seek to positively persuade organisations to adopt Travel Plans or as disincentives (sticks), whereby organisations are penalised for not introducing or operating a Travel Plan. Secondly, the Travel Plan support measures are arranged in terms of resource (money, time, expertise, political effort, legislative effort) required, where the greater the resource needed the darker the shading.

Here it should be noted that in general terms there would seem to be a close relationship between resource expended and the effectiveness of the intervention in terms of public policy outcomes delivered. In addition, the importance of appropriate indirect measures needs to be reiterated once again. Bluntly, even the most resource intensive Travel Plan programme measures can be significantly compromised if wider public polices relating to the transport system (and the car in particular) are not consistent with the goals of the Travel Plan policy agenda.

Overall, this chapter has largely assumed a progression in the level of support offered by public authorities and the wider Travel Plan industry – from a minimal to a relatively comprehensive level of intervention. And, as can be seen there have been occasions where some of the more robust approaches can work well. However, as discussed in earlier Chapters 'support' is only one aspect of development in which Travel Plans need to be progressed.

Consequently, Chapter 7 will draw together the lessons learnt particularly from Chapters 3, 4, 5 and 6 in order to propose a series of recommendations for practitioners and policy makers that will hopefully lead to Travel Plans being more effective in the future.

Table 6.7 Travel Plan support: resource intensity and effectiveness

Travel Plan incentives	Travel Plan disincentives
Information provision	
Capacity building – network creation	
Capacity building – outsourcing	
Subsidy provision	Regulation
Fiscal incentives	Fiscal penalties

A crucial role for local authorities is to help Travel Planners to network with each other

Dropping off passengers by private car near airport terminals is now not possible at most UK airports following a terror attack at Glasgow Airport in June 2007. Such 'trigger' events can significantly affect the viability of a Travel Plan

References

BBC. 2010. Hospital parking charges to stay, government says, 16 September. [Online] Available at http://www.bbc.co.uk [accessed 11 April 2011].

BEST Workplaces for Commuters. 2005. *Commuter Tax Benefits: Implementing Commuter Benefits as one of the Nation's Best Workplaces for Commuters*. Research Report for Projects 68-W6-0029 and 2W-0693-NBLX, Office of Air and Radiation, (Washington DC, US Environmental Protection Agency).

BEST Workplaces for Commuters. 2011. *BEST Workplaces for Commuters*. [Online] Available at http://www.bestworkplaces.org [accessed 18 April 2011].

Budd, L., Griggs, S. Howarth, D. and Ison, S. G. 2011. A fiasco of volcanic proportions? Eyjafjallajökull and the closure of European airspace, *Mobilities*, 6(1), 31–40.

Cairns, S., Sloman, L., Newson, C., Anable, J., Kirkbride, A. and Goodwin, P. 2005. *Making Smarter Choices Work*, (London: Department for Transport).

Comsis Corporation. 2002. *Public Agency Guidance on Employer-based TDM Programs*. Prepared for Transit Cooperative Research Program Project B-4, TCRP Web Document 22, Contractor's Final Reports, December. (Silver Spring: Comsis Corporation).

Department for Transport (Great Britain). 2008. *Ruth Kelly Launches* £140 *Million Cycling Fund*, Press release, (London: Department for Transport), 21 January.

Department of the Environment, Transport and the Regions (Great Britain). 1998. *A New Deal for Transport: Better for Everyone*, (London: The Stationery Office).

Dill, J. 1998. Mandatory employer-based trip reduction, *Transportation Research Record*, 1618, 103–110.

DirectGov (Great Britain). 2011. *Climate Change and Protecting the Environment*. [Online] Available at http://www.actonco2.direct.gov.uk [accessed on 11 April 2011].

Enoch, M. P. 2002. UK parking cash out experience, and lessons from California, *Traffic Engineering and Control*, 43(5), 184–187.

Enoch, M. P. and Ison, S. G. 2006. Levying charges on private parking: lessons from existing practice, *World Transport Policy and Practice*, 12(1), 5–14.

Enoch, M. P. and Ison, S. G. 2011. Travel plans: a way forward? *Urban Design and Planning*, forthcoming.

Enoch, M. P., Zhang, L. and Morris, D. 2005. *Organisational Structures for Implementing Travel Plans: A Review*, Research Report for OPTIMUM2 Project, London Borough of Southwark Cluster, European Regional Development Fund (Lille: EC-NWE-Interreg IIIB Programme).

Enoch, M. P., Zhang, L. and Ison, S. G. 2007. Unlocking the potential of site based mobility management through local travel plan groups, *World Transport Policy and Practice*, 13(2), 23–39.

European Platform on Mobility Management. 2011. *What is EPOMM Plus?* [Online] Available at http://www.epomm.eu [accessed 9 April 2011].

Glaister, S. 2002. UK transport policy 1997–2001, *Oxford Review of Economic Policy*, 18(2), 154–186.

Hendricks, S. 2008. Four challenges to incorporating Transportation Demand Management into the land development process, *Transportation Research Record*, 2046, 30–36.

Highways Agency (Great Britain). 2011. *Tackling Congestion by Influencing Travel Behaviour*. [Online] Available at http://www.highways.gov.uk [accessed 5 May 2011].

iBase Systems. 2008. *iTrace*. [Online] Available at http://itrace.org.uk/ [accessed 18 April 2011].

Kadesh, E. and Roach, W. T. 1997. Commute trip reduction: a collaborative approach, *Energy Policy*, 25, (14–15), 1217 1225.

Liftshare. 2011. *Liftshare.com: Better Together*. [Online] Available at http://www.liftshare.com [accessed 8 April 2011].

Litman, T. (1997) *Employer Provided Transit Passes: A Tax Exempt Benefit* (Victoria, BC: Victoria Transport Policy Institute).

MAX. 2009a. *MaxLupo Guidelines for Integrating Planning and Mobility Management*, Successful Travel Awareness Campaigns and Mobility Management Strategies, Project 518368, (Brussels: EC Sixth Framework for Research and Technological Development). [Online] Available at http://www.max-success.eu [accessed 8 April 2011].

MAX. 2009b. *MAXimise Mobility Management: A Guide to Results from the MAX-project – Aiming to Extend, Standardise and Improve Mobility Management,* Successful Travel Awareness Campaigns and Mobility Management Strategies, Project 518368, (Brussels: EC Sixth Framework for Research and Technological Development). [Online] Available at http://www.max-success.eu [accessed 8 April 2011].

National Business Travel Network. 2011. *Homepage*. [Online] Available at http://www.nbtn.org.uk [accessed 6 April 2011].

National Cooperative Highway Research Program. 2010. *State Department of Transportation Role in the Implementation of Transportation Demand Management Programs*. Research Results Digest 348, (Washington DC: Transportation Research Board).

Organisation for Economic Cooperation and Development and the International Transport Forum. 2010a. *Effective Transport Policies for Corporate Mobility Management*, Transport Research Centre, (Paris: OECD).

Organisation for Economic Cooperation and Development and the International Transport Forum. 2010b. *Trends in the Transport Sector 1970–2008*, Transport Research Centre, (Paris: OECD).

Pagano, A. M. and Verdin, J. 1997. Employee trip reduction without government mandates: cost and effectiveness estimates from Chicago, *Transportation Research Record*, 1598, 43–48.

Potter, S., Enoch, M. P., Rye, T., Black, C. and Ubbels, B. 2006. The tax treatment of employer commuting support: an international review, *Transport Reviews*, 26(2), 221–237.

Rye, T. 1999. Employer transport plans – a case for regulation, *Transport Reviews*, 19(1), 13–31.

Rye, T. 2002. Travel plans: do they work? *Transport Policy*, 9(4), 287–298.

Rye, T., Green, C. Young, E. and Ison, S. G. 2011. Using the land-use planning process to secure travel plans: an assessment of progress in England to date, *Journal of Transport Geography*, 19(2), 235–243.

Shoup, D. 2005. *The High Cost of Free Parking*, (Chicago: American Planning Association).

Taxsaver.ie. 2010. *Everything you Need to Know about Taxsaver*, Information Booklet, (Dublin: Taxsaver.ie). [Online] Available at http://www.taxsaver.ie [accessed 5 April 2011].

Transport for London. 2011. *Publications*. [Online] Available at http://www.tfl.gov.uk/corporate/projectsandschemes/workplacetravelplanning/7680.aspx [accessed 18 April 2011].

Tyler, S., Enoch, M. P. and Zhang, L. 2007. Setting up local travel plan groups: the future of workplace travel planning in urban conurbations? Examples from London, UK. Paper presented to the 86th Annual Meeting, *Transportation Research Board*, Paper 07-2164, Session 553: New International Research in Transportation Demand and Parking Management, Washington DC, 21–25 January.

Ubbels, B., Enoch, M. P., Potter, S. and Nijkamp, P. 2004. *Unfare Solutions*, (London: Spon).

Van-Ristell, J. A. 2011 Investigating policy impacts on school travel, PhD Thesis, (Loughborough University: School of Civil and Building Engineering).

Washington State Department of Transport. 2011. *Commute Trip Reduction (CTR)*. [Online] Available at http://www.wsdot.wa.gov/TDM/CTR/default.htm [accessed 18 April 2011].

Washington State Energy Office. 1995. *1995 Report to the Washington State Legislature*, (Olympia: Washington State Energy Office).

Chapter 7
Recommendations for Practice and Policy and Implications for the Future

1 Introduction

The purpose of this chapter is to present the conclusions reached over the course of this book so far; to use these to lead into a series of recommendations the practice and policy as to how to more effectively apply the Travel Plan concept; and to explore what this might mean for Travel Plans worldwide in the future. In identifying the available actions, some of the more general points that arise from Chapter 2 are reported first, followed by a taxonomy of possible measures that has been derived based on the findings arrived at in the four preceding chapters, which looked at Travel Plans from segmental, scale, structural and support perspectives. Subsequent sections report these recommendations,[1] explore possible pathways to implementation and summarise the recommendations by stakeholder, namely Travel Planning organisation, Travel Plan industry, and government. Finally, data on the status of Travel Plans currently throughout the world is presented, followed by conclusions as to the likely future of the Travel Plan concept.

2 Overall Travel Plan Concept

At this point it is useful to draw some more general conclusions and recommendations from the material in Chapter 2.

- First, the lack of a suitable term for the 'Travel Plan' concept is a major hindrance to its role in transport policy (and indeed in devising an appropriate title for this book), as evidenced by the wide array of terms used over the years and throughout the world. Suffice to say that 'site-based TDM' is clumsy; 'mobility management' is already too associated with mobility impaired and social mobility usages; 'smarter choices' is so vague as to be meaningless; and 'Employer Travel Plans' conjures images of workplace summer trips to the seaside. Yet a suggested 'better' alternative has so far not come to mind.

---

1 Note: Some recommendations are repeated in different sections due to their cross-disciplinary nature.

- Second, it is clear that whilst there is some variation in what a Travel Plan is in each of its manifestations, fundamentally they comprise a means of delivering measures to help facilitate sustainable transport through an additional 'mediating institution'.
- Third, Travel Plans can be highly effective at meeting sustainable transport objectives at the organisational level in specific circumstances.
- Fourth, Travel Plans tend to be far less effective at meeting sustainable transport objectives at the network level, due primarily to their adoption not being sufficiently universal.

Resulting from these conclusions, the following recommendations emerge:

- There is a need for a suitable and universally understood term to be developed to replace 'Travel Plan'.
- The underlying Travel Plan concept then needs to be communicated more widely and more effectively not only to potential implementing organisations, but also to the transport sector.
- Efforts need to be taken to increase the prevalence of the 'special circumstances' necessary for highly effective Travel Plans to thrive.
- Travel Plans need to be made far more attractive to the mediating institutions to increase levels of take up.

3 Segment-related Measures

Chapter 3 reviews the evidence as to the effectiveness of Travel Plans by segment, and concludes that some trip types and destinations seem to provide more scope for effective Travel Planning activities than others.

Specifically, it suggests that business trips, supplier trips and emergency patient trips to hospital, commuting and education trips are suitable targets for Travel Plans, whilst visitor, leisure customer and non-emergency hospital patient trips present more of a challenge to Travel Planners – with shopping trips falling somewhere in the middle depending on circumstances. With this in mind, in general, the most appropriate contexts for Travel Plans would seem to be (very large) public sector employers and transport interchanges, followed by (large) private sector employers, schools and (large scale) planned events. Less suitable are attractions and retail segments, and least suitable for Travel Plans would be residential sites (although it should be noted that there are exceptions). In addition, it is worth considering Travel Plans in certain circumstances for the 'education' factor. School Travel Plans, for example, are deemed valuable because they help influence the attitudes of children in how they see the car. And, in a similar way, Event Travel Plans (for anything from the Olympic Games to long-term road works) have been shown to 'persuade' or trigger car users to try other modes due to special circumstances which they probably otherwise would not have done.

From this, it is recommended that:

- Where resources are limited, effort should first be targeted at those destination-types most suitable for Travel Plans, namely public sector employers, transport interchanges, private sector employers, schools and (large scale) planned events.
- Within these destination types, it is suggested that effort should first be targeted at business trips, supplier trips, emergency patient trips to hospital, commuting and education trips where resources are limited.
- However, it is important to recognise that other destination and trip types may also be appropriate for Travel Plan interventions in special circumstances.
- Travel Plans in schools and for events potentially have a wider 'education' function in persuading die-hard car users to try alternatives, and so could be worth focusing on for this reason.

4 Scale-related Measures

From the perspective of scale, Chapter 4 reports that, in general, the larger the Travel Plan organisation the greater the potential for Travel Plans to be implemented and to achieve their objectives. This is because larger employers are more likely to have a manager in a position whose role could include implementing Travel Plan measures, and can generate a large potential market for those measures once adopted. On the other hand though, there are some issues with increasing scale. Thus, smaller employers have fewer layers of decision-making authority to negotiate; are less likely to occupy multiple sites which can be a complicating factor; and face lower overall costs when administering a Travel Plan (TCRP 2003). As a result, one mechanism for trying to achieve the benefits of both a larger scale and smaller scale Travel Plan, is to establish a local Travel Plan group or network of some form.

Such Local Travel Plan Networks (LTPNs) range from informal, loosely defined networks drawing on minimal resources to more formal and legally defined organisations with access to a series of possible promotional, regulatory and fiscal tools. They can be either transport-focused or more generally aligned and are heavily influenced by contextual factors such as the primary actors, the motivations for group formation, the perceived scope and scale of the problem and the organisational environments, along with geographic, political and institutional factors. In terms of performance, there appears to be a direct correlation between the type of LTPN (and the corresponding level of resources available) and the contribution the network can make to reducing transport externalities. There is also an underlying story that LTPNs have not yet realised their full potential either at the individual local network level or at the wider community transport system level.

Accordingly, in order to take account of the scale factor in Travel Planning, it is recommended that:

- Where resources are limited, large organisations should be targeted first, followed by smaller organisations that can conveniently be formed into some form of Local Travel Plan Network.
- When preparing a LTPN, practitioners should devise a good strategic plan and idea, consider the site context, generate a strong network of members and wider stakeholders and focus attention on how to fund the network.
- When implementing a LTPN, practitioners should aim for ‘quick and easy wins’, pay attention to marketing efforts and tailor the solutions to the needs of the LTPN.
- When operating a LTPN, practitioners should maintain the stakeholder network and funding streams and measure and report results.
- Local policy makers should pursue LTPNs only where they are an appropriate measure and consider LTPNs as early as possible during the planning process for new developments and when designing related transport and wider policy measures. In addition, they should initially provide strong leadership, but then try and increasingly involve and empower the stakeholders once the network is up and running.
- Nationally, policy makers need to act as LTPN ‘champions’ to ‘spread the word’, continually assess LTPN benefits and regularly reinforce the LTPN concept through policy actions.
- At the supra-national level, the role of policy makers should be to promote LTPNs as good practice in terms of transport and wider policy goals, establish funding opportunities to kickstart networks and conduct research and disseminate results.

5 Structural Measures

The structural perspective of Travel Plans is investigated in Chapter 5. This finds that Travel Plans currently very much reflect their local authority heritage as a means of addressing transport problems which means that they are usually only adopted by organisations with a transport-related problem that needs to be addressed – a niche market – in spite of Travel Plans delivering sometimes significant benefits both to the organisations and to society more generally. The key lesson is that Travel Plans will only become mainstream if they are adopted more widely by organisations, but this will only happen if the Travel Plan ‘product’ is modified to take account of the fact that these organisations view the world rather differently to local authorities and the Travel Plan industry. Thus, for a ‘step change’ in take up to be delivered then an additional ‘re-oriented’ Travel Plan model should be developed.

Specifically, it is recommended that:

- Re-oriented Travel Plans should focus on the needs of organisations more than those of transport policy makers. In other words, Travel Plans should become predominantly business management rather than public policy tools;
- Re-oriented Travel Plans should consist of 'easy to implement' individualised measures that can be incorporated into existing management practices such as ISO 14001 and EMAS rather than a holistic 'package' of measures that are typically applied as an additional function;
- Re-oriented Travel Plans should be implemented ideally by organisations incentivised to do so, rather than by being required to do so;
- Re-oriented Travel Plans should be implemented at a time that is convenient to the organisation rather than as and when a local authority has the necessary resources;
- Re-oriented Travel Plans should be promoted by faces familiar to organisations (for example management consultants, environmental experts) rather than by transport consultants with whom an organisation is not likely to have a pre-existing relationship.

6 Support Measures

The following recommendations are derived from Chapter 6, and are arranged in sections concerned with conclusions and recommendations on information, capacity building, subsidy, regulation, fiscal, institutional, indirect measures and leadership issues.

6.1 Information Measures

Whilst government departments often have strong records is in the delivery of practical advice and research on the use and performance of Travel Plans, knowledge of the Travel Plan concept – even within the transport sector, never mind in the wider world – remains limited, and so much work remains to be done in publicising Travel Plans more generally. From this, the information-focused recommendations that arise are that:

- Greater efforts are required to raise awareness of Travel Plans at the most basic level amongst organisations.
- (Even more) work needs to be done on disseminating examples of Travel Plan best practice amongst organisations.
- The benefits of supporting Travel Plans need to be made more obvious, and if necessary should be enhanced by the provision of additional support from governments and local authorities. Such enhancements may include:

financial grants and greater access to transport policy decision makers (see later).

- In selling the benefits, information needs to be focused far more on the needs of the 'client' organisations, both in terms of style and content, rather than on the public policy benefits sought as currently. For example, some organisations require evidence of how corporate social responsibility goals could be attained, while others will focus on environmental issues or efficiency savings.

6.2 Capacity Building Measures

Strongly related to information provision, is the need for the Travel Plan industry to be developed, a process that comprises several elements. In the short-term, this involves encouraging organisations to adopt Travel Plans; establishing support networks; and outsourcing Travel Plan tasks. And in the longer term, the challenge is to develop a core of professionals in the area to embed these short-term measures into organisation management and public policy frameworks and thus take Travel Planning on to the next level through education and training. In building Travel Plan capacity, then, the recommendations are that:

- There is a need to identify policy champions able to drive the Travel Planning agenda across the transport sector and beyond.
- Best practice experience should be shared more effectively through new research projects, Travel Plan industry networks and also through wider business and industrial networks as well as though trade unions, public sector bodies and the public transport sector. In particular, more effort needs to be made to incorporate business travel and freight transport within the auspices of Workplace Travel Planning, while lessons should be learnt from the School Travel Plan and individualised marketing (personal journey planning) sectors.
- In terms of outsourcing, there is scope for a range of services provided by consultants, public transport operators, local authorities and so on, offering everything from 'do-it-yourself kits' of Travel Plan tools, through 'off-the-shelf' measures and on to specially tailored 'bespoke' site-specific Travel Plans. These are required at every level – from the individual via the organisational, the neighbourhood and the local to the regional. Crucially, they would need to be tailored to meet the motivations and needs of organisations, and should therefore be designed in such a way as to tap into the various motivations behind organisational decision-making processes – for example, improving profits, reducing costs, addressing transport issues and meeting Corporate Social Responsibility and/or environmental targets.
- Other related products that should be encouraged include home-shopping schemes; local network enabled joint distribution and home delivery

schemes; videoconferencing; neighbourhood work centres; organisation-sponsored (and hence heavily discounted) public transport tickets; car club-based carpooling schemes; and shared/pool bicycle schemes.
- Existing training and education programmes need to be stepped up to increase the pool of Travel Planners equipped with the necessary tools for the role.
- Finally, the profile of the Travel Planning profession needs to be raised by giving attention to the skill level of Travel Plan coordinators, possibly with bursaries being made available by local authorities and/or governments to encourage further study on sustainable transport and Travel Plan training and education programmes.

6.3 Subsidy Measures

The lack of sufficient financial resources, and revenue funding in particular, is a significant barrier to the advancement of Travel Plans, and subsidy schemes potentially provide a means of addressing this. Subsidy recommendations are that:

- On funding, Travel Plans would benefit from a higher proportion of revenue rather than capital funding being made available.
- Mechanisms for targeting subsidies in specific 'Travel Plan friendly' locations should be considered so as to maximise any benefits.
- Subsidies for 'capacity building' the Travel Planning sector would be highly beneficial.
- Subsidies for developing Travel Plan infrastructure should be considered.
- Possible future finance sources such as developer fees should be investigated to pay for subsidy programmes.
- Non-cash subsidies (perhaps in the form of time savings) could be offered to 'reward' Travel Plan accredited organisations to encourage wider participation in Travel Planning.

6.4 Regulation Measures

There are several benefits to adopting regulations to promote Travel Plans, such as being able to affect all employers in an area who meet all pre-determined criteria; covering large areas; and securing economies of scale in providing services to groups of employers. However, there are also problems which include administrative costs; the perceived burden on and so resentment from employers; difficulty of regulating lots of employers without a large bureaucracy; and the problem of how to regulate small employers. On balance then, the recommendations that emerge are:

- Where none currently exists, legislation and/or regulations legally requiring existing 'appropriate' organisations to introduce Travel Plans should be explored.
- If a mandatory approach to Travel Plans is adopted, then the need for a robust framework for monitoring the effectiveness of Travel Plans should be explored to support this.
- The concept of an accreditation system for Travel Plans should be examined as a means of ensuring minimum standards and of increasing their effectiveness. This would range from a very simple exercise for small organisations to something more tailored and involved for larger bodies. Ideally, it should be tied in to existing environmental management system or corporate social responsibility regimes.
- Any adopted Travel Plan accreditation system should be time limited to ensure continued development.
- Travel Plans should (continue to) be linked to the planning process.
- Travel Plans need to be considered at a far earlier stage in the planning process if they are to realise their full potential.
- Local authorities should ensure that the costs of monitoring and enforcing a Travel Plan are 'built in' to the original submission of each planning proposal where monitoring will be required.

6.5 Fiscal Measures

Potentially the most effective way of persuading organisations to adopt Travel Plans is to offer them financial incentives for using 'sustainable transport' options and/or to penalise them through disincentives for using a car through various aspects of tax regimes to do so. So far, the most common mechanisms have been through incentives through the personal tax system, though other approaches (corporate taxation, road user charging schemes, parking taxes) can also be applied.

Recommendations here are that:

- There is a need for governments to revisit tax systems as they relate to Travel Plans. In particular, they should consider using not only the personal tax system, but wider options too (for example, corporate tax, purchase taxes, road user charges, parking taxes).
- Governments need to continue exploring ways of making Travel Plan benefits increasingly attractive *vis-à-vis* car-based benefits.
- Governments need to clarify existing Travel Plan-related tax benefits, which are currently often too complicated for organisations to understand.
- Governments need to publicise existing (and any future) Travel Plan-related tax benefits far more effectively.

6.6 Institutional Measures

In order to implement the previous recommendations effectively, it is necessary to consider the institutional framework. This is particularly relevant in the case of Travel Plans, given the whole concept centres on Travel Plans being the most appropriate mode of delivery for a range of other transport policies. In other words, large organisations or perhaps groups of smaller organisations may often be the most effective means of delivering parking, walking, cycling or public transport policy measures than a regional or local authority – providing it has the necessary resources and support. Recommendations here therefore are that:

- The role of governments with respect to Travel Plans needs to be clarified. Fundamentally, a decision needs to be taken as to the future of Travel Plans. Are they to continue to evolve slowly as currently on minimal resources, or are they to potentially become the core component in delivering the TDM agenda? A half way house may be to test a highly involved approach, perhaps through the use of TDM demonstration sites (see later).
- Governments should consider how to organise the strategic application of Travel Plans and Local Travel Plan Networks in a way that would allow them to become a major policy delivery mechanism of TDM measures and resultant benefits on the ground.
- Travel Plans should only be seriously pursued where the organisational and site contexts make them an appropriate measure.
- Implementation efforts should be focused on those organisations in the locations that would demonstrate the best returns.
- Local authorities serious about pushing the TDM agenda should first introduce a Travel Plan for their own premises, if one does not already exist.
- Local authorities should encourage and support the establishment of Local Travel Plan Networks where conditions are suitable. LTPNs can provide significant economies of scale and can reduce barriers to entry. They can be a particularly effective way of involving small and medium-sized enterprises in Travel Planning.
- A network of Travel Plan policy champions is required to promote the concept more effectively both in existing (transport-related) arenas of policy and practice, but also in those of other sectors (business, management, CSR, environmental management and so on).
- Efforts to share best practice experience through new research projects, Travel Plan industry networks and wider networks such as employer, trade union and public transport member organisations need to be continued and stepped up.
- Lessons can be learnt from the business travel and freight sectors, and from the adoption of School Travel Plans in terms of bringing them more into the mainstream.

- Governments should consider establishing 'Centres of Excellence' (CoEs) within local authorities to deliver TDM measures. Here, one or more local authority, public transport operator, consultant, the regional Highways Agency office would work together to implement TDM measures through Travel Plans and/or Local Travel Plan Networks. They would provide advice and develop a series of Travel Plan services for local organisations with travel issues; help establish Local Travel Plan Networks; carry out monitoring and enforcement roles; and develop and prepare bodies of evidence of how well Travel Plans are working. The centres would be eligible for government financing and would benefit from greater flexibility in how they met their local transport planning process targets.
- Advice and support also needs to be made available for organisations located in non-Centre of Excellence areas that wish to develop Travel Plan initiatives. This could be introduced on either a national or regional basis.
- The development and adoption of an agreed accredited Travel Plan standard, particularly if it can be integrated with other accreditation management programmes and internationally applied, would be a positive step.

6.7 Indirect Support Measures

A core finding from the book relates to the need for wider supporting measures to 'lock in' the benefits afforded by Travel Plan interventions. The following recommendations are suggested for consideration.

- Governments should seek to better integrate Travel Plans within transport, planning and wider government policy proposals.
- Travel Plans can help support other transport policy interventions to be more effective.
- Specifically, the status of Travel Plans needs to be raised in local transport planning processes.
- There is a need to benchmark Travel Plan initiatives against other transport policy initiatives, for example park-and-ride to determine their relative effectiveness in meeting policy goals.

6.8 Leadership Measures

Overall, probably the most urgent need is for governments to develop strategies setting out exactly how Travel Planning fits in to their overall transport and wider public policy agendas. Beyond this, governments need to communicate the importance of Travel Planning to local authorities and other agencies including public transport operators and other transport suppliers, multi-national organisations, public sector organisations, business groups and trade unions.

Proposed recommendations resulting from this are that:

- Action should be taken to ensure Travel Plans become as effective as possible.
- The priority is for governments to develop strategies setting out how they see the future of Travel Plans.
- The message that Travel Plans are an important and integral part of a government's transport, planning (and wider) policy needs to be highlighted perhaps through high level meetings between ministers and senior managers of agencies such as transport suppliers, multi-national organisations, public sector organisations, business groups and trade unions.
- Efforts should be made by government representatives to 'persuade' large organisations with a large number of sites across the country to adopt Travel Plans as company policy.
- Governments need to work on convincing local authorities and other agencies that potentially comprise the Travel Plan industry that Travel Plans are a core element of its transport and planning policy arsenal.
- The Travel Plan concept should be re-launched – but only once it is known how they are to be applied. This should incorporate the adoption of the accredited Travel Plan as a minimal standard.
- Strategic lessons should be learnt from the experiences of 'good practice' agencies in terms of implementing a strategy for Travel Plans.
- There is a need to identify key policy champions, able to drive the Travel Plan agenda, and to increase capacity building efforts for the Travel Plan sector.

7 Potential Future Policy Pathways

From the above policy proposals, it is suggested that governments could follow three possible pathways with respect to Travel Plans. These could be termed Business as Usual, Centre of Excellence and Comprehensive Local Transport Delivery Tool. In addition to these government-focused approaches it is important to note a parallel strand that cuts across the three pathways and emerges from the discussion in Chapter 5, namely the degree to which the re-oriented Travel Plan model is adopted as opposed to the traditional model. Finally, it is useful to consider just how these pathways might be introduced in practice.

7.1 Business as Usual (BAU)

The business as usual model would see minor modifications to regulations, institutions and the tax system. For example, government could:

- Raise Travel Planning with organisation heads through Senior Ministers to encourage them to directly raise the profile of Travel Planning within their companies.

- Encourage home-shopping schemes; local network enabled joint distribution and home delivery schemes; videoconferencing; neighbourhood work centres; organisation-sponsored (and hence heavily discounted) public transport tickets; car club-based carpooling schemes; and shared/pool bicycle schemes.
- Encourage local authorities to consider Travel Plans earlier in the planning process.
- Grant tax relief on the amount of money they spent on Travel Plan measures through Enhanced Capital Allowances or Industrial Building Allowances, or maybe allow Travel Plan Tax Credits.
- Encourage local authorities to support the establishment of Local Travel Plan Networks where conditions are suitable.
- Continue to encourage the sharing of best practice experience through new research projects, Travel Plan industry networks and wider networks such as employer, trade union and public transport member organisations.
- Incorporate business travel and freight transport within the auspices of workplace Travel Planning.
- Link Special Parking Areas or Parking Districts – areas where parking offences have been decriminalised and are now enforced by the local authority – more closely with Travel Plans.
- Explicitly include Travel Plan elements such as the cost of car parking and business travel in auditing procedures for local authorities, government departments and public sector bodies.
- Benchmark Travel Plan performance against other transport policy mechanisms, ideally by using the same (albeit suitably modified) assessment criteria as used to judge more traditional transport planning, highway and/or railway infrastructure schemes.

Such measures would arise in an ad hoc manner as a result of actions by local authorities or other organisations. There would be no real push from government.

From a performance perspective, the BAU model would likely see only a relatively modest shift in the uptake of Travel Plan measures and only in local authority areas that are already convinced by the TDM agenda. As a result, while the costs of implementing such improvements would remain relatively small so would the corresponding benefits.

7.2 Centre of Excellence (COE)

The centre of excellence model would see governments recognising the benefits of Travel Plans and then investing significant resources in a tightly focused way in specific geographical areas known as TDM Centres of Excellence. Some resources would also be made available to 'sympathetic' neighbouring authorities for specific Travel Plan projects. Minor regulatory changes would be made to strengthen planning rules, and the tax regime would be significantly adjusted to favour non-car modes, while institutional and policy measures would see Travel Plans becoming a key element. However, there would be no real movement on indirect support measures. Specific policy measures here for governments might be to:

- Develop a strategy, setting out exactly how Travel Planning fits in to an overall Transport Department agenda, and for government as a whole.
- Set up TDM Centres of Excellence (TDMCoE).
- Set up TDM Accredited Authorities (TDMAA).
- Reward local authorities that are taking a lead on TDM with greater political and policy support such as, improved access to funds (along with greater flexibility in spending them).
- Encourage local authorities to require Travel Plans for all developments as a requirement for obtaining planning permission in TDMCoEs and TDMAAs.
- Allow local authorities to require Travel Plans for all organisations within specifically designated areas in TDMCoEs and TDMAAs.
- Establish some form of Innovation Fund for Travel Plan measures, available only to TDMCoEs and TDMAAs.
- Reintroduce Travel Plan Bursary scheme for those authorities contemplating becoming a Centre of Excellence for TDMCoEs and TDMAAs.
- Reintroduce a modified Site-specific Advice (SSA) programme for use by TDMCoEs and TDMAAs.
- Provide grants for Travel Plan related infrastructure, Local Travel Plan Networks, or innovative Travel Plan related tools more generally, for example joint distribution centres and delivery clubs for networks of retail outlets, car clubs, video-conferencing suites or neighbourhood 'virtual work centres' in TDMCoEs and TDMAAs.
- Consider allowing TDMCoE and TDMAA local authorities to offer discounts with the proposed introduction of a Supplementary Business Rate for organisations with accredited Travel Plans.
- Consider tax benefits for employees living near to their workplaces, tax benefits for employers providing accommodation for their employees (within a limited distance of the work site), tax breaks for communication equipment such as tele-conferencing and video-conferencing suites, neighbourhood work centres and so on, designed to minimise the need for people to commute and make business trips.
- Encourage home-shopping schemes; local network enabled joint distribution and home delivery schemes; videoconferencing; neighbourhood work centres; organisation-sponsored (and hence heavily discounted) public transport tickets; car club-based carpooling schemes; and shared/pool bicycle schemes.
- Support the expansion of research into Travel Planning issues and training opportunities for Travel Planners.

- Conduct wider government campaigns to educate the public as to the threat of global warming; and the benefits of using alternatives to the car.
- Link Special Parking Areas or Parking Districts more closely with Travel Plans.
- Modify project appraisal criteria for transport planning, highway and railway schemes to include the potential role of Travel Plan (and wider TDM) measures where appropriate.
- Explicitly include Travel Plan elements such as the cost of car parking and business travel in auditing procedures for local authorities, government departments and public sector bodies.
- Consider Travel Plans when introducing parking, environmental and planning regulations, and require local authorities to do the same.
- Benchmark Travel Plan performance against other transport policy mechanisms, ideally by using the same (albeit suitably modified) assessment criteria as used to judge more traditional transport planning, highway and/or railway infrastructure schemes.

The key element here is that local authorities would make the decision as to the role of Travel Plans in their own areas, but with far more support than is currently available.

Based on the GORS (2007) report findings cited earlier, in the UK context roughly one quarter of local authorities made significant reference to the Smarter Choice agenda in their Local Transport Plans (five-year transport policy plans) while just over half made a reasonable reference. Realistically, therefore, it could be expected that a significant proportion of these might seek to increase their commitment (and hence deliver a step change in transport and wider societal benefits).

Once again, the level of costs incurred would be commensurate to the benefits.

7.3 Comprehensive Local Transport Delivery Tool

Finally, the Comprehensive Local Transport Delivery Tool would see (area-based) Travel Plans becoming the key instrument on a national scale for delivering the most localised mobility management measures, such as walking and cycling improvements, parking policy, and so on. Here, every traffic generator in the country would become responsible for their transport impacts. Regulations would thus be introduced (perhaps through local Environmental Laws) requiring accredited Travel Plans to be in place at each site. Subsidies would also be made available to each Travel Plan and pro-Travel Plan tax incentives developed to ‘sweeten the pill’. Concurrently, government would implement a whole range of indirect support measures, for example introducing road user charging either nationally, or in more places at a local level would strongly reinforce the role of the Travel Plan or re-applying VAT to parking spaces.

Such an approach would see transport policy (and crucially spending) being reoriented such that mobility management measures supported where necessary by capacity improvements became the default option, rather than the other way around as at present.

While the benefits could be substantial for a relatively minor financial cost, the institutional barriers to be overcome should not be under-estimated and would require a major political investment to convince first the transport sector itself and then the wider public to its merits.

7.4 The Traditional and Re-oriented Travel Plan Models

Cutting across the future pathways discussed is the issue of whether Travel Plans remain predominantly traditional; whether the re-oriented model will become the dominant model; or whether a blend of types emerges. In the opinion of the author it is the latter alternative that is most likely. This is because traditional Travel Plans clearly work for those organisations with transport problems, whilst the re-oriented model would effectively broaden the appeal of the Travel Plan concept beyond the current (fairly narrow) niche it occupies currently.

In any event, many of the recommendations proposed would benefit both models equally. However, there is a key difference in the philosophy behind each approach and this has implications in terms of the balance of the recommendations. Specifically, under the traditional Travel Plan model the 'ideal' support measures centre on requiring or pushing organisations to adopt a tailored, transport-focused package of measures. By contrast, the re-oriented model seeks to provide organisations with small-scale 'off-the-shelf' organisation focused services and so the support required is more aimed at creating an environment that encourages the adoption of measures.

In other words, the nature of the Travel Plan type directly affects whether the role of the local and national government and the wider Travel Plan industry – that is, the degree to which this role is as a regulator/enforcer/funder (under the traditional model), and the degree to which it is a facilitator/enabler/service provider (under the re-oriented model).

7.5 Realising the Future Pathways

Of course it is one thing to propose alternative pathways to improving the take up and effectiveness of Travel Plans in theory, but another thing entirely to see the plan successfully carried out in practice. Hence, Gunn (1978) notes the difficulties faced when implementing public policy and hence the importance of thinking about implementation in advance. This will need to take account of the specific context and should finalise exactly not only what measures to implement; but also consider when; how (should it be incrementally applied or adopted in one go? What legislation should it use?); by whom (which bodies will take responsibility for which aspects?); where (in what jurisdiction?); and why (what are the motivations for these measures?).

In addition, such a plan might also explore if there is any scope for enhancing the chances of a 'successful' implementation through the judicious use of implementation strategies. For example, Enoch et al (2004) looks at several cases where controversial transport schemes were implemented with the help of techniques such as 'compensating losers'; 'bribing' key interest groups to change behaviour; 'highlighting the benefits' of the new scheme; 'offering [the individual and/or organisation] increased choice' generated by the new scheme; offering a policy as being the 'lesser of two evils'; stating that 'it might have been worse' if the policy had not been implemented; 'adapting tried and tested methods' to minimise the impact of the new scheme; and using the 'Trojan Horse' – that is implementing a transport measure as a solution to another issue entirely which is less controversial.

8 Recommendations by Stakeholder Type

Table 7.1 now pulls together the recommendations made throughout this chapter, and outlines the stakeholders best placed to implement them.

From Table 7.1 it can be seen that the majority of recommendations are for national and local government action – either alone or in partnership with the Travel Plan industry – whilst for others the emphasis is on the Travel Plan industry to lead.

Overall, then, transport issues will always be an extremely difficult issue for any government to address given the scale of the challenge to be faced and the integral part transport plays in people's lives, not just from a practical perspective but also from financial, economic, environmental, social and particularly political perspectives. Specifically, traditional transport supply-side solutions involve large commitments of financial and political investment over a long period of time, and negatively affect significant numbers of people while they are being introduced. Consequently, the temptation for decision makers has been to leave them to one side. Unfortunately, though it is an entirely understandable policy, this 'do-nothing' approach has seen transport conditions continue to worsen and in the medium to long-term is simply not sustainable by any definition.

Travel Plans offer a possible way of avoiding some of these unattractive traits by seeking to manage the use of the existing transport system more efficiently. They do this by targeting 'easy wins' – certain journey types in particular locations and at specific times – that result in the maximum benefit for the least cost, by convincing people that they will benefit by sometimes changing their travel behaviour when provided with a range of usable alternatives that are designed to suit their specific needs. Whilst this is by no means an easy task, an increasing body of evidence is showing that better information coupled with a mix of incentives and disincentives can deliver significant transport and societal benefits in a reasonable timeframe

Table 7.1 Summary of recommendations for practice and policy

Type of measure	Recommendation	Actor(s) affected
Overall Travel Plan concept	• There is a need for a suitable and universally understood term to be developed to replace 'Travel Plan'. • The underlying concept needs to be communicated not only to potential implementing organisations, but also to the transport sector. • Efforts need to be taken to increase the prevalence of the 'special circumstances' necessary for highly effective Travel Plans to thrive. • Travel plans need to be made far more attractive to the mediating institutions to increase levels of take up.	
Segment	• Resources should first be targeted at destination-types most suitable for Travel Plans. • Resources should be targeted at most suitable trip types. • Note, other destination and trip types may also be appropriate in special cases. • The 'wider 'education' role of school and event Travel Plans in persuading car users to try alternatives mean they may be worth focusing on.	Local government, and Travel Plan industry
Scale	• Large organisations should be targeted first, followed by smaller organisations that can form a local Travel Plan network. • When preparing a LTPN, practitioners should devise a strategic plan and idea, consider the site context, generate a strong network of members and wider stakeholders, and focus attention on how to fund the network. • When implementing a LTPN, practitioners should seek 'quick and easy wins', pay attention to marketing efforts, and tailor the solutions. • When operating a LTPN, practitioners should maintain the stakeholder network and funding streams, and measure and report results.	Local government and Travel Plan industry
	• Local policy makers should pursue LTPNs only where appropriate measure and consider LTPNs as early as possible during the planning process for new developments and when designing related transport and wider policy measures. They should initially provide strong leadership, but increasingly involve and empower the stakeholders once network is up and running.	Local government
	• Nationally, policy makers should be LTPN 'champions', assess LTPN benefits, and reinforce the LTPN concept through policy actions.	National government
	• At the supra-national level, policy should promote LTPNs, establish funding opportunities, conduct research and disseminate results.	Supra-national government
Structure	• Re-oriented Travel Plans should focus on organisational needs and be business management tools. • Re-oriented Travel Plans should comprise 'easy to implement' measures that can be incorporated into existing management practices. • Re-oriented Travel Plans should be implemented ideally by organisations incentivised to do so. • Re-oriented Travel Plans should be implemented at a time that is convenient to the organisation. • Re-oriented Travel Plans should be promoted by faces familiar to organisations.	Travel Plan industry
Support – information	• Greater efforts are required to raise awareness of Travel Plans. • Travel Plan best practice needs to be more widely disseminated. • Travel Plan benefits need to be made more obvious to organisations. • Travel Plan benefits need to be made more obvious to government.	National government and Travel Plan industry

Table 7.1 *Continued*

Type of measure	Recommendation	Actor(s) affected
Support – capacity building	• Policy champions need identifying to drive the Travel Planning agenda. • Best practice experience should be shared more effectively through new research projects, Travel Plan industry and wider networks. • There is scope for a range of Travel Plan services to be outsourced to consultants, public transport operators, local authorities. • Other related products that should be encouraged include home-shopping schemes and local joint distribution and home delivery schemes. • Existing training and education programmes need to be stepped up to increase the pool of Travel Planners. • Finally, the profile of the Travel Planning profession needs to be raised.	Travel Plan industry and national government
Support – subsidy	• On funding, Travel Plans would benefit from a higher proportion of revenue rather than capital funding. • Mechanisms for targeting subsidies in specific 'Travel Plan friendly' locations should be considered. • Subsidies for 'capacity building' the Travel Planning sector would be highly beneficial. • Subsidies for developing Travel Plan infrastructure should be considered. • Possible future finance sources such as developer fees should be investigated to pay for subsidy programmes. • Non-cash subsidies (perhaps in the form of time savings) could be offered to 'reward' Travel Plan accredited organisations.	National and local government
Support – regulation	• Where none currently exists, legislation/regulations legally requiring existing organisations to introduce Travel Plans should be explored. • If a mandatory approach is adopted, then the need for a robust framework for monitoring the effectiveness of Travel Plans should be explored. • An accreditation system for Travel Plans should be examined to ensure minimum standards and increase effectiveness. • Any adopted Travel Plan accreditation system should be time limited. • Travel plans should (continue to) be linked to the planning process. • Travel plans need to be considered at a far earlier stage in the planning process. • Local authorities should ensure that the costs of monitoring and enforcing a Travel Plan are 'built in' to planning proposal submissions.	National and local government
Support – fiscal	• Governments should revisit tax systems as they relate to Travel Plans. • Governments should continue to explore how to make Travel Plan benefits more attractive *vis a vis* car-based benefits. • Governments need to clarify existing Travel Plan-related tax benefits. • Governments need to publicise existing (and any future) Travel Plan-related tax benefits far more effectively.	National government

Table 7.1 *Concluded*

Type of measure	Recommendation	Actor(s) affected
Support – institutional	• The role of governments with respect to Travel Plans needs to be clarified. • Governments should consider how to better organise the strategic application of Travel Plans and local Travel Plan networks.	National government
	• Travel Plans should only be pursued where the organisational and site contexts make them appropriate. • Implementation efforts should focus on organisations at suitable sites. • Local authorities should introduce a Travel Plan for their own premises. • Local authorities should help introduce local Travel Plan networks where conditions are suitable.	Local government
	• Policy champions are required to promote Travel Plans both in existing (transport-related) arenas of policy and practice, but also in those of other sectors.	Travel Plan industry
	• Efforts to share best practice experience need to be continued and stepped up.	National government
	• Lessons can be learnt from the business travel, freight and school Travel Plan sectors.	Travel Plan industry
	• Governments should establish 'Centres of Excellence' (CoEs) within local authorities to deliver TDM. • Advice and support also needs to be made available for organisations located in non-Centre of Excellence areas that wish to develop Travel Plan initiatives.	National government
	• An agreed accredited Travel Plan standard would be a positive step.	Travel Plan industry
Support – indirect	• Governments should better integrate Travel Plans within transport, planning and wider government policy proposals. • Governments should better integrate and support TDM measures within transport, planning and wider government policy proposals. • The status of Travel Plans needs to be raised in local transport planning processes. • Travel Plan initiatives should be benchmarked against other transport policy initiatives.	National and local government
Support – leadership	• Action is needed to ensure Travel Plans become effective. • The priority is for governments to develop a vision for the future of Travel Plans. • The message that Travel Plans are important to government needs to be highlighted. • Government representatives should 'persuade' large organisations to adopt Travel Plans as company policy. • Governments need to convince local authorities and other agencies that Travel Plans are important.	National and local government
	• The Travel Plan concept should be re-launched – but only once it is known how they are to be applied. • Strategic lessons should be learnt from the experiences of 'good practice' agencies. • Key policy champions, able to drive the Travel Plan agenda need to be identified.	Travel Plan industry, national and local government

and at an affordable financial (and political) price. Moreover, by involving organisations with whom individuals already have an established relationship (for example employer, school, hospital, retail or leisure facility provider), there is often more of a willingness to become involved in such a process.

Crucially, though, to be most effective such an approach needs to be implemented as part of a comprehensive overall package which includes a range of complementary transport and wider measures, and in particular a major marketing role to convince citizens (and organisations) that such a policy is in their best interest not just in the short-term but in the longer term too. In short, it requires leadership at the national level to kick start what would be a paradigm shift in transport policy so that transport providers, local authorities and other key stakeholders were fully aware of the new regime, how and what it seeks to achieve, and why.

In determining the likelihood of such a paradigm shift occurring, it is first useful to identify the status of Travel Planning across the world at the moment, before speculating on what the future of the concept might be.

9 Travel Plans Now

In determining the current status of Travel Plans currently in European countries, the first section draws extensively from the 2009 *Mobility management monitors* 'country profiles' of the European Platform on Mobility Management website (EPOMM 2009) and presents an interpretation of these results in Table 7.2.

Table 7.2 Travel Plan issues, actions and actors

Country	Mobility management context	Current status of Travel Plans	Future prospects
Austria	Pilot programmes set up in mid-1990s, then national Klimat Activ programme from 2004.	Advanced for companies, schools and events.	Carbon agenda will see MM become more important.
Finland	MM piloted from 2002.	Not much progress as yet with Travel Plans – concept to be launched soon.	Too early to tell.
France	First Travel Plan group in 1998; legal recognition in 2000 and slow dissemination from mid-2000s.	Travel Plans not yet widely disseminated.	Largely positive.
Netherlands	MM adopted early 1990s. Regulations then tax support tried. Covenant approach since 2005.	Strong organisation focus and minimal link to transport benefits. Now compulsory for large road schemes.	Outlook more positive than for many years. Recent step change in interest perhaps due to threat of regulation.
Portugal	Mobility centres in mid-1990s but closed down later. Recent revival of interest by organisation (mid-2000s).	Travel Plans in a few large organisations.	Slow progress expected.
Sweden	MM measures in place since mid-1990s.	Travel Plans for organisations now emerging.	Slow progress expected.

UK	MM measures since mid-1990s.	Travel Plans formally recognised since 1998 and steady dissemination since. Good progress with school Travel Plans.	Unclear what future holds.
Belgium	Mobility Covenants from 1996, and second generation in place from 2009 in Flanders.	Good progress with School Travel Plans. Company Travel Plans less widely disseminated.	Outlook more positive than for many years.
Bulgaria	First MM initiatives from mid-2000s due to EU projects.	Limited progress.	Too early to tell.
Czech Republic	Little evidence of 'new' MM activity.	Bus services have always been in place for schools and for factories.	Business as usual.
Estonia	First MM initiatives from mid-2000s due to EU projects.	Limited progress.	Too early to tell.
Greece	Little evidence of MM activity.	No evidence of Travel Plans.	Business as usual.
Hungary	First MM initiatives from mid-2000s due to EU projects.	Bus services have always been in place for schools and for workplaces.	Too early to tell though interest is growing.
Italy	MM initiatives from mid-late 1990s.	Travel Plans legally required from 1998, so relatively widespread.	Barriers restricting Travel Plan development that are unlikely to be addressed soon.
Lithuania	Little evidence of MM activity though there have been EU projects.	Little application of Travel Plans beyond EU projects.	Too early to tell.
Poland	First MM initiatives from mid-2000s due to EU projects.	Bus services have always been in place for schools and for workplaces.	Will require efforts from 'bottom-up'. Too early to tell.
Romania	First MM initiatives from mid-2000s due to EU projects.	Little application of Travel Plans beyond EU projects.	Business as usual.
Slovenia	Government action from 2006/7 in recognising MM.	Little application of Travel Plans beyond EU projects.	Too early to tell.
Spain	National MM policy since 2005 though not really followed.	Significant subsidy nationally for Workplace Travel Plans, and regulations supporting TPs in Catalonia since late 2000s.	Too early to tell though government looks to be pushing MM.
Denmark	MM policy only since 2008, though strongly supportive of non-car modes previously.	Not really a feature previously but now being launched.	Outlook fairly positive, with many wider supporting car restraint measures already in place.
Ireland†	MM policy in Dublin since mid-1990s.	Tax breaks for green commuting in place since early 2000s. National Travel Plan strategy now underway.	Economic recession possible threat to Travel Plan development.

Source: Based on EPOMM (2009) and interviews with national experts.[2]
Note: † Not based on EPOMM (2009) report.

From Table 7.2 it is possible to group the countries into six categories based on their mobility management/Travel Plan status.

2 Please also refer to Chapter 2 for more details on Travel Plan development in the Netherlands, UK and Italy.

- Of the countries reported in Table 8.1, Greece is the only nation not to have actively considered MM.
- Next, Finland, Bulgaria, Estonia, Lithuania, Romania and Slovenia have only just begun to make the first steps towards MM and Travel Plans. European projects seem to have been important in these efforts.
- Portugal and Sweden are making a little more progress with MM and Travel Plans specifically, while the Czech Republic, Hungary and Poland still have strong elements of company and school MM in the form of works buses from the days prior to mass car ownership.
- National governments in Austria, France, Spain, Denmark and Ireland have each provided support for MM programmes in the form of subsidies, regulations, tax breaks and/or supporting transport infrastructure and services, but Travel Plans are a relatively recent concept and are not yet common.
- The UK, Belgium and Italy comprise the next grouping. Here Travel Plans have some of the same support mechanisms mentioned previously, but in addition have been in place for longer and so are more widespread.
- Lastly, the Netherlands is currently the nation with the most developed Travel Plan programme in Europe. In particular, the recent threat of national legislation to require organisations to adopt Travel Plans coupled with a new emphasis on developing tools to help address organisation-focused goals in addition to offering support through other mechanisms is felt to be significant.

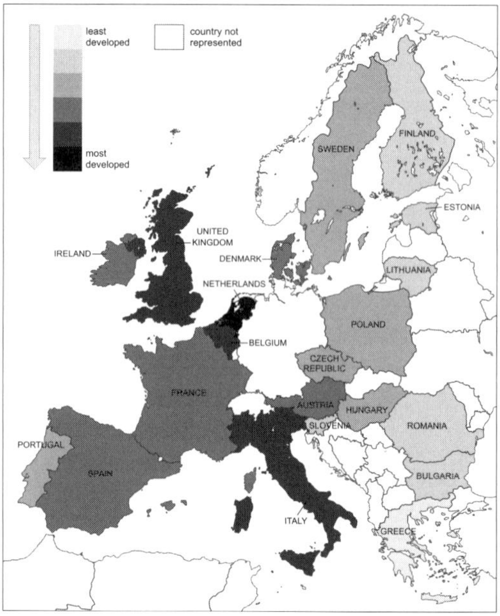

Figure 7.1 The current status of Travel Plans across Europe

Figure 7.1 illustrates this map of current Travel Plan status across Europe.

One European country not covered above is Germany. This is partly due mobility management being administered not at the national but at the regional and local levels, meaning that in general results can best be described as 'patchy', and partly because (perhaps as a consequence) there seems to have been relatively little interest in Travel Plans. Thus, as noted in Chapter 3 the exceptions are that there has been a long involvement in Car Free Development and in Personalised Journey Planning in specific areas, whilst Workplace Travel Plans for example seem to have been confined to regions such as North Rhine Westphalia, and cities like Munich and Frankfurt.

Elsewhere, in the United States Travel Plans have been in evidence since the early 1970s (as noted in Chapter 2) and so are relatively mature. On the ground the dominant model is whereby travel-related services are provided to an organisation either by a form of Local Travel Plan Network, local/regional authority or a private sector supplier. In policy support terms, information, subsidy, regulation and fiscal mechanisms have all been tested at different times and in different locations, although indirect and leadership measures have often not been as strong as they might have been. One outcome of these factors is that Travel Plans are now so integrated into business practices that in the research arena at least they seem to be largely ignored by researchers. Regarding the future then, it is likely that Travel Plans in the USA have now diffused as far as they can be without a significant boost perhaps from circumstances and/or government. However, the fear in the USA is that the oil price rises of the 1970s (which started Travel Plans in their current guise and initially generated a significant private sector industry), then fell again a few years later which took away the market as people went back to using their cars and so bankrupted many of the new TDM companies, and so there is a residual cautiousness about making the same mistake again.

Next, Travel Plan schemes in Canada were only initiated from the mid-1990s, two of the earliest being at the University of British Columbia in 1997 and Nortel Networks in Ottawa, Ontario in 1998 (Noxon Associates and Computing Solutions 2008). Interestingly, the emphasis has largely been on following the US model whereby tools such as Employer Transit Passes (annual public transport tickets sold to employers at a significant discount for all of their staff and handed out to employees as a benefit) and ride matching (carpooling) schemes rather than packages of measures form the basis of the concept (Lewinson, S. 2011. Personal communication, President, ACT Canada, 23 February). Regional and local authority support has tended to focus on information, policy and subsidy support.

Meanwhile, in Australia, the first Travel Plan is understood to have been implemented at the University of New South Wales in 1995, although the first programme of measures was implemented a year later in Western Australia (Wake, D. 2011, Personal communication, Project Officer, TravelSmart Workplace, Department of Transport, Western Australia, 22 February). Subsequently, Travel Plans were adopted in New South Wales, Victoria, South Australia and most recently Queensland (Meiklejohn, D. 2011. Personal communication, Consultant,

UrbanTrans (ANZ), 24 February). In addition, the Northern Territory and Australian Capital Territory ran Federally-funded schemes which have since been discontinued, whilst Tasmania has not had any involvement with the concept. Meiklejohn notes that most states have followed the British Travel Plan model, but that Access Melbourne, a Transportation Management Association did try a less structured (more US) approach with some success whilst the State of Victoria has also been involved in developing Travel Plan tools for businesses to use 'off-the-shelf'. Overall, however, Travel Planning has never enjoyed significant levels of funding and the future of the concept as a whole remains uncertain, with the role of Travel Plan programmes currently being reviewed in several states.

In neighbouring New Zealand, the first Workplace Travel Plan was set up in Auckland at a Vodafone site in 2004 (Baker, L., Haizelden, P. and Philbin, S. 2011. Personal communication, Consultants, Opus International Consultants, Auckland, 4 May). This was followed by the introduction of a Travel Plan process in the region between 2005 and 2010 which aimed to introduce a number of Travel Plans and saw Travel Plan networks established at large public and private sector organisations and free tools such as travel to work surveys, marketing material, carpool database and Personalised Travel Planning developed for employers, as well as technical guidance for those new to Travel Plans and partial funding for consultants to help design and implement them. However, re-structuring of the local government structure in the area in 2010 subsequently stalled the process and it has yet to restart. Meanwhile, at the federal level, the election of the National Party to Government in November 2008 has also impacted on Travel Plans, due to a policy focus on construction rather than TDM. That said, there are some local examples of cities that continue to develop TDM initiatives. Thus, in Hamilton Travel Plans are mandatory for large new developments requiring planning permission; and in Hastings, a Model Communities initiative to promote walking and cycling will see Travel Plans used to support the infrastructure developments along specific corridors.

Mobility management in Japan has tended to focus on the individualised Travel Feedback Programmes approach, although Ayako Taniguchi notes that workplace Travel Plans have started to become more commonplace in the last year or two (Taniguchi, A. 2011. Personal communication, Lecturer, Tsukuba University, Tsukuba, 23 February). She notes that [the lack of] funding for such schemes is a major issue.

Finally, the author is aware of some practitioners requesting information on Travel Plans for possible application in several less developed countries including Bangladesh, India and China, where huge pressures being put on transport systems would seemingly make Travel Plans worthy of consideration.

Drawing on the data discussed, Figure 7.2 illustrates how Travel Planning has developed in a number of countries from around the world. Clearly this is not an exact science,[3] but nevertheless it does at least give some indication as to the story of Travel Plans and their current status in terms of how well embedded they are in the respective policy frameworks. In particular it can be seen how sensitive Travel Plans are to external forces and especially to the policy support measures available in each nation state.

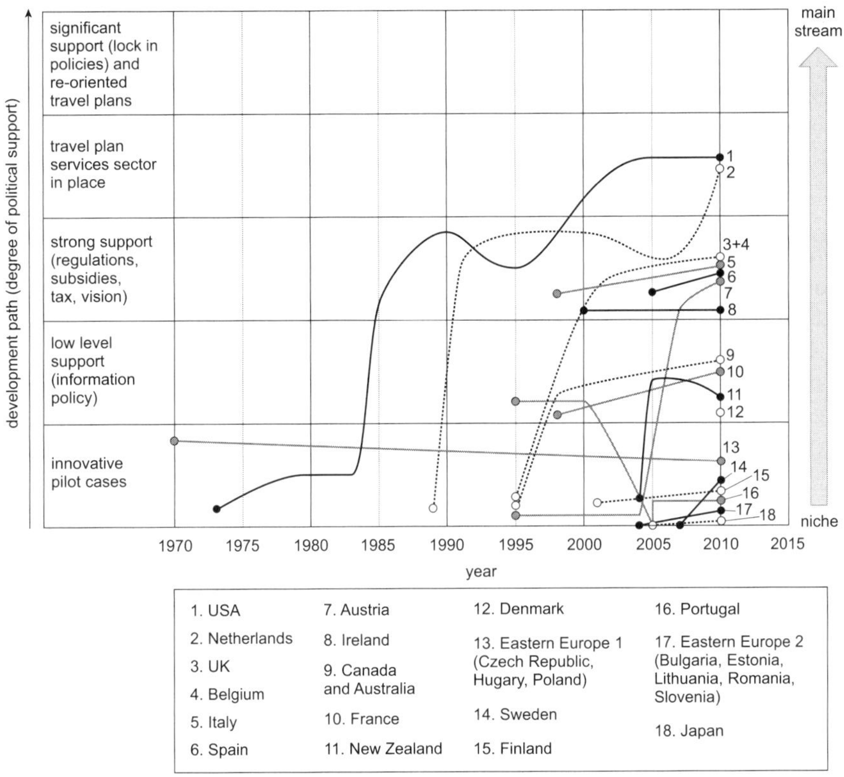

Figure 7.2 Travel Plans and their evolution in selected countries

10 The Future Potential of Travel Plans

Looking to the future, Table 7.2 and the additional evidence presented in the same section also indicate that with regard to their likely development, the picture is rather mixed. Thus, some countries/provinces look set to see the role of Travel

3 In particular, because the 'developmental stage' categories on the vertical axis are amalgams of measures that are difficult to plot; because assigning 'scores' across a country is extremely difficult in many cases because of the huge range of experiences between regional and local areas; and because the data is largely anecdotal.

Plans expanding such that over time they may well become a core element of local transport policy and practice either from their being fairly well established (as in the Netherlands, Austria and Belgium), or from closer to a standing start (as in France, Portugal, Sweden, Spain and Denmark). Next, in a number of other countries the impact of Travel Plans looks set to remain as before, either because the process has become static (Italy), lost direction (UK), not yet really started (Finland), or because Travel Plans have not yet even made the policy agenda (Czech Republic, Romania and particularly Greece). Finally, in countries like Bulgaria, Estonia, Hungary, Lithuania, Poland and Slovenia, the future appears somewhat less clear cut because it is simply far too early in the process to tell.

Seeing these experiences as part of the bigger picture, transport problems – congestion, air pollution, energy use, land take, traffic collisions and so on – are not likely to go away. Indeed, as vehicle ownership and use continue inexorably to rise towards there being two billion of them worldwide by 2020 (at least half of which will be cars) (Sperling and Gordon 2009), if anything these problems are likely to significantly worsen. In addition, a range of issues related to car use (including continued fluctuations in oil prices due to instability in the Middle East and the 'peak oil' issue; the challenge of global warming; and concerns about children's health being affected by loss of mobility and air pollutants) add to the pressure for a change in how the transport system functions. Moreover, traditional 'supply-side' solutions such as road building and providing better public transport are becoming increasingly more difficult to introduce, due to there being less money available and due to increased resistance by the public to the idea of destroying communities and natural habitats. Meanwhile, the alternative demand-side measures (like road pricing and controls on parking) remain politically difficult to sell to populations that are becoming more and more dependent on the use of the car.

Based on these observations, it would seem that the broader contextual environment for Travel Plans is becoming an increasingly benign one for them to develop and grow. However, for this to be fully taken advantage of, there really needs to be a step change in how Travel Plans are designed, marketed, and supported by the Travel Plan industry and by governments of all levels.

In summary, therefore, it is clear that Travel Plans potentially offer a quick, inexpensive, politically acceptable and effective mechanism managing transport and for addressing a whole range of transport issues. Yet currently, despite some exceptions as noted previously, in most locations across the world the likelihood of such a scenario occurring probably remains slim. This is due mainly to ignorance of the concept and its benefits and to the Travel Plan not yet being sufficiently adapted to the needs of the organisations that are required to deliver it, which given the acknowledged difficulties in dealing with the 'transport mess' is rather frustrating.

It was to challenge this position that this book was written. Specifically, the ideas put forward here attempt to offer practitioners and policy makers a road map in considering how to make the most of the Travel Plan approach, and hopefully therefore help to lower the odds in favour of a more economically, socially and environmentally sustainable transport future developing. It remains to be seen whether such a map can help influence the changes in direction required for those

There is a need for more training and education for practitioners if Travel Plans are to be mainstreamed

As the Bedzed development in the London Borough of Sutton illustrates, limiting the number of car parking spaces through the planning process can help generate interest in Travel Planning solutions

charged with navigating the transport policy course, though one can hope that it will do so, at least to some extent.

References

Enoch, M. P., Wixey, S. and Ison, S. G. 2004. Practical lessons for winning support for radical transport proposals, *World Transport Policy and Practice*, 10(1), 34–40.

European Platform on Mobility Management. 2009. *Mobility Management Monitors, 2009 country profile reports*, Deliverables for the EPOMM PLUS Project (Brussels: EC-EACI-IEE-STEER Programme), November. [Online] Available online at http://www.epomm.eu [accessed 2 May 2011].

Government Operational Research Service (Great Britain) and Department for Transport (Great Britain). 2007. *Review of the Take-up of Smarter Choices in Local Transport Plans*, (London: Government Operational Research Unit Report).

Gunn, L. A. 1978. Why is implementation so difficult? *Management Services in Government*, 33, 169–176.

Noxon Associates and Computing Solutions. 2008. *The Case for TDM in Canada: Transportation Demand Management Initiatives and their Benefits*, (Ottawa: Association for Commuter Transportation of Canada).

Sperling, D. and Gordon, D. 2009. *Two Billion Cars: Driving Toward Sustainability*, (Oxford: Oxford University Press).

Transit Cooperative Research Program. 2003. *Strategies for Increasing the Effectiveness of Commuter Benefits Programs*, TCRP Report 87, (Washington DC: Transportation Research Board).

Index